GEOMORPHOLOGY
OF
DESERT DUNES

Sand dunes cover up to a quarter of many desert regions. Determining the factors that determine their shape, size and dynamics is important to an understanding of many aspects of the geomorphology and ecology of desert regions, as well as to resource management in these areas.

Recent developments in satellite images of deserts, orbital images of Mars and Venus, and oil and gas discoveries in ancient dunes, as well as detailed studies of surface processes, have significantly advanced our understanding of desert environments and raised questions about arid geomorphology.

Geomorphology of Desert Dunes offers a comprehensive understanding of how desert dunes are formed, how they change, and their environmental importance. Analysing dune types, patterns, sand seas and sediments, and dune dynamics and processes at different temporal and spatial scales, the author draws on extensive research from the deserts of Africa, North America, India and Australia.

Investigating the role of climatic change, the book concludes that a better understanding of dune processes and dynamics is vital for effective and appropriate avoidance and mitigation of environmental problems in arid regions.

Invaluable for students and research workers in geomorphology and sedimentology, this book will also be of value to resource managers, ecologists and engineers with an interest in desert regions.

Nicholas Lancaster is a Research Professor at the Desert Research Institute, University and Community College System of Nevada.

A VOLUME IN THE
ROUTLEDGE PHYSICAL ENVIRONMENT SERIES

Edited by Keith Richards
University of Cambridge

The Routledge Physical Environment series presents authoritative reviews of significant issues in physical geography and the environmental sciences. The series aims to become a complete text library, covering physical themes, specific environments, environmental change, policy and management as well as developments in methodology, techniques and philosophy.

Other titles in the series:

ENVIRONMENTAL HAZARDS: ASSESSING RISK AND REDUCING DISASTER
K. Smith

WATER RESOURCES IN THE ARID REALM
E. Anderson and C. Agnew

ICE AGE EARTH: LATE QUATERNARY GEOLOGY AND CLIMATE
A. Dawson

MOUNTAIN WEATHER AND CLIMATE, 2ND EDITION
R.G. Barry

SOILS AND ENVIRONMENT
S. Ellis and A. Mellor

Forthcoming:

GLACIATED LANDSCAPES
M. Sharp

HUMID TROPICAL ENVIRONMENTS AND LANDSCAPES
R. Walsh

PROCESS, ENVIRONMENT AND LANDFORMS: APPROACHES TO GEOMORPHOLOGY
K. Richards

To Judith

whose support and assistance made it all possible

CONTENTS

CONTENTS

FIGURES

PLATES

TABLES

PREFACE

This book is a personal view of the geomorphology of desert dunes that has developed from my research on dune sediments, processes, and dynamics over the past two decades. This work has been made possible by research support from a variety of organizations, including the Royal Geographical Society, the Royal Society, and the National Geographic Society; the Universities of Cambridge, Witwatersrand, and Cape Town; the Transvaal Museum; the Foundation for Research Development (South Africa); NASA, NATO, and the National Science Foundation.

In the course of my work, I have interacted with many individuals who have provided assistance and discussion of problems and results. I thank them all for their company and time. I would like to single out A. T. (Dick) Grove (University of Cambridge), Mary Seely (Desert Ecological Research Unit of Namibia), Ron Greeley (Arizona State University), Gary Kocurek (University of Texas at Austin), and Bill Nickling (University of Guelph) as individuals who have made major personal contributions to my thinking on aeolian processes and dunes.

Many individuals have helped me in producing this book. In particular, Dale Ritter of the Quaternary Sciences Center, Desert Research Institute, provided institutional support; Rob Bamford did all the drafting for the figures; and Liane Alessi assisted with proof reading.

ACKNOWLEDGEMENTS

I thank the following individuals and organizations for permission to reproduce illustrations.

Academic Press Inc.: Figures 2.1, 2.16, 2.17, 3.9, 3.17, 4.3, 8.12.
American Association for the Advancement of Science: Figure 2.7.
American Geographical Society: Figure 6.11.
American Society of Civil Engineers: Figures 6.20 and 6.21.
Robert Anderson: Figure 2.7.
Edward Arnold: Figures 3.15 and 4.17.
A. A. Balkema Publishers: Figures 3.11, 4.2, 4.13, 4.15, 4.17, 4.19, 5.19, 5.20, 5.33, 6.2, 6.10, 6.23, 7.12, 7.13, 7.14, 7.19, 8.1.
B. T. Batsford Publishers: Figure 4.29a.
Belhaven Press: Figures 1.1 and 7.1.
Blackwell Scientific Publications: Figures 2.20, 3.7, 4.6, 5.16, 5.21, 5.22, 5.24, 5.26, 5.30, 5.31, 5.34, 6.7, 6.8, 7.3, 7.5, 8.11, 8.14.
Cambridge University Press: Figures 2.4 and 5.2.
Chapman and Hall: Figures 2.6, 2.10, 2.12, 2.13, 2.20, 4.7, 5.3, 5.5, 5.8, 5.27, 5.32 and 6.2, 6.5 (part).
Elsevier Science Publishers: Figures 4.1, 4.4, 4.5, 4.11, 4.12, 6.13, 6.19, 8.2, 8.3, 8.7, 8.8, 8.9, 8.10.
Gebrüder Borntraeger: Figure 6.5b.
Geographical Association: Figure 6.7.
Geographical Society of New South Wales: Figure 3.14.
Ron Greeley: Figures 2.4, 5.2.
J. Hardisty: Figures 2.8, 2.18.
Kluwer Academic Publishers: Figures 6.16, 7.4.
Macmillan Magazines: Figures 2.8, 2.18, 6.9.
Methuen & Co: Figure 2.4.
Royal Meteorological Society: Figure 5.6.
Scandinavian University Press: Figure 5.25.
SEPM (Society for Sedimentary Geology): Figures 4.8, 4.9, 4.14, 5.4, 7.2, 7.9, 7.18.

ACKNOWLEDGEMENTS

Springer Verlag: Figure 8.15.
Robert Wasson: Figure 6.9.
John Wiley & Sons: Figures 2.11, 2.15, 5.28, 5.29, 7.7.

1

INTRODUCTION

Quartz sand dominates aeolian deposits in most arid regions, as dust-sized particles are carried out of the desert entirely or are trapped by vegetation or rough surfaces on desert margins (Tsoar and Pye 1987). The greater part (> 95 per cent) of this sand occurs in accumulations known as sand seas or ergs (Wilson 1973) that comprise areas of dunes of varying morphological types and sizes, as well as areas of sand sheets. Smaller dune areas are known informally as dune fields. Major sand seas occur in the eastern hemisphere arid zones of the Sahara, Arabia, central Asia, Australia and southern Africa, where they cover as much as 45 per cent of the area classified as arid (Figure 1.1). In North and South America there are no large sand seas and dunes cover less than 1 per cent of the arid zone.

PROBLEMS AND CHALLENGES IN DUNE GEOMORPHOLOGY

Plate 1 shows a Landsat image of the Namib sand sea, in southwestern Africa. Even with a ground resolution of 80 m per pixel, this image shows a wealth of information about dune form and patterns, and raises many fundamental questions about the geomorphology of the area.

Dominating the image are large south-to-north oriented linear dunes. Why are these dunes oriented in this direction, and what factors determine their size and very regular spacing? What determines the very different linear dune orientation to the east of the sand sea? Dune type changes from crescentic through linear to star dunes from the coast to the inland edge of the sand sea. Why does this occur, and what factors determine dune type? Other questions immediately arise. What are the relations between the dunes and the river valleys that enter the sand sea from the east and how has the sand sea reacted to changes in sea level and climate? A closer aerial view of some of the coastal crescentic dunes (Plate 2) reveals further questions: why are there two sets of dunes: the main crescentic ridges and the smaller superimposed dunes?

On the same scale, the Landsat image of the Gran Desierto sand sea in northern Mexico (Plate 3) shows a much smaller sand sea, but one with a very

1

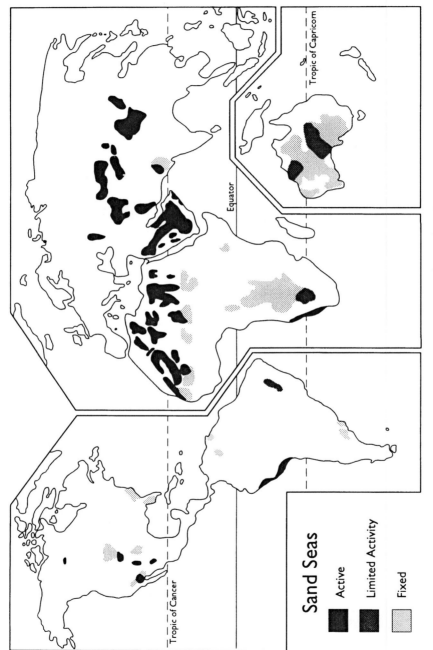

Sand Seas

■ Active

■ Limited Activity

▨ Fixed

Figure 1.1 Location of major sand seas and dune fields (after Thomas 1989).

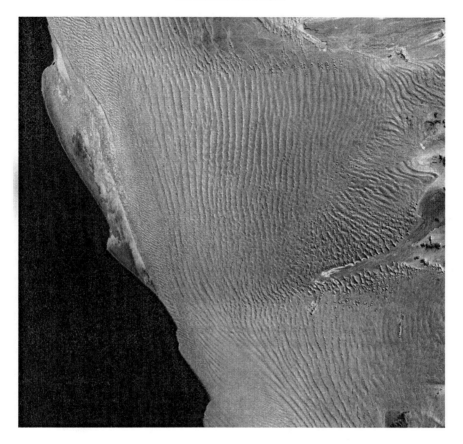

Plate 1 Landsat image of the central parts of the Namib Sand Sea.

different set of dune patterns. Unlike the Namib sand sea, the Gran Desierto shows a series of discrete areas of different dune types. Again, the questions of controlling factors and formative processes arise, but with the additional consideration: why does this sand sea appear different from the Namib? Why do large star dunes occur next to small crescentic dunes? Is this an effect of dune size and process–form interactions? Do different dune areas represent multiple generations of dunes formed at different times and in different conditions, and therefore does the climatic, tectonic, and sea level history of this area play a role in determining dune patterns? What is the role of different sediment source areas and how have these changed over time?

These images point to some of the challenges in dune geomorphology today. Whereas we have quite a good knowledge of the main aspects of dune

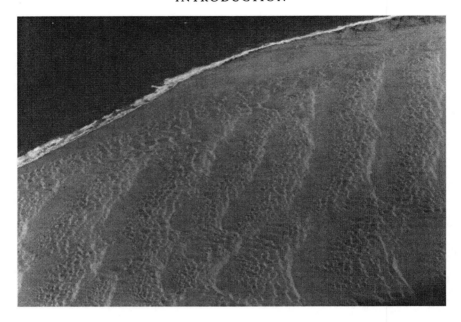

Plate 2 Oblique aerial view of compound crescentic dunes in the Namib Sand Sea.
Note two sets of dunes: the main forms and the superimposed forms.

morphology and are beginning to understand the processes that maintain
different dune types, we still are a long way from understanding the
fundamental controls of dune size and spacing, and the processes by which
dunes and sand seas develop.

DEVELOPMENT OF MODERN DUNE STUDIES

Early work on desert dunes was dominated by descriptions of dune form,
based largely on exploratory investigations. Important contributions were
made by Hedin (1903) in central Asia, Gautier (1935) and Newbold (1924)
in the Sahara, Blandford (1876) in the Indian subcontinent, Thesiger (1949)
in Arabia, and Passarge (1904) in the Kalahari. Following this exploratory
phase, scientific investigations of desert regions by Capot-Rey (1947),
Monod (1958), Beadnell (1910), and Bagnold (1933) in the Sahara; Madigan
(1946) in Australia; Kaiser (1926) in the Namib, and Hack (1941) in the
United States, produced a wealth of important information on the occurrence of dunes and their basic forms. Attempts to relate form to process were
very few. One exception is the pioneering work of Cornish (1914), but it was
not until the seminal work of R. A. Bagnold (1941) in the Egyptian desert
that the relations between the characteristics of the surface wind and sand

4

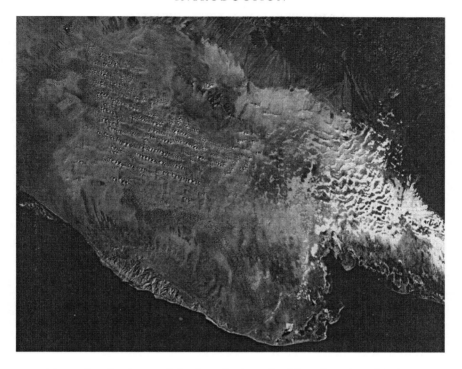

Plate 3 Landsat image of the Gran Desierto Sand Sea of northern Mexico.

transport processes emerged and were quantified.

In recent years, the focus of studies of dunes has changed radically. Three main developments have stimulated investigations of dunes: (1) Satellite images of desert regions have become widely available, focusing attention on the development of dune patterns and their relations to sediment sources and wind regimes; (2) Orbital images of Mars (Mariner and Viking missions) (Greeley, Lancaster *et al*. 1992) and Venus (Magellan) (Greeley, Arvidson *et al*. 1992) have shown that aeolian processes occur on other terrestrial planets, and have resulted in sharpened concepts of the fundamentals of the physics of sediment transport by the wind; and (3) The discovery of oil and gas in Mesozoic sandstone in the western United States and northwest Europe has stimulated a series of detailed studies of aeolian sediments and sedimentary processes.

INTRODUCTION

ASPECTS OF DUNE STUDIES

Studies of dunes are concentrated in three main areas: (1) the description of dune forms and patterns, (2) analyses of dune sediments, and (3) investigations of dune dynamics and processes.

Description of dune forms and patterns

The starting point for all aspects of desert dune geomorphology is the identification and description of different dune types. Initially, descriptions of dunes were made in the course of ground investigations, often during the exploration of desert regions. Names for the forms (e.g. *seif, silk, barchan, zibar*, etc.) were derived from the rich terminology for desert landforms employed by the local population (Bagnold 1951; Breed *et al.* 1979). Later investigations utilized aerial photography to great advantage in their descriptions (e.g. Wilson 1972). As a result, data on dune patterns and trends became more widely available. In the 1970s, the availability of satellite images of desert regions made it possible to study the patterns of dunes throughout sand seas with relative ease and led to the realization that basic dune forms in different sand seas are remarkably similar (McKee 1979a). The realization that dunes of similar characteristics occur in widely separated sand seas has focused attention upon the factors which control their morphology and morphometry.

Studies of dune sediments

Studies of dune sediments have concentrated upon investigations of grain shape, colour and mineralogy; grain size and sorting characteristics; and sedimentary structures in dune sediments. Many investigations of dune sediments have been linked to attempts to positively identify sands in the rock record as aeolian and to characterize their depositional environments (Ahlbrandt 1979). This has frequently involved comparisons of aeolian sands with those deposited in marine, coastal, fluvial, or glacial environments (Mason and Folk 1958). However, recent work has shown that the use of textural parameters (grain size, sorting) is unreliable as an indicator of depositional environments. Consequently, attention has turned to studies of the variability of grain size and sorting parameters of dune sands over individual dunes and within sand seas and dune fields, as well as to the analysis of primary sedimentary structures.

Studies of sedimentary structures in dunes were pioneered by McKee and have provided much valuable information on the ways in which dunes accumulate (McKee 1966). However, the logistics of carrying out such studies in most desert regions have prevented their full potential from being realized. Recent work has been concerned with the identification and

6

description of small-scale sedimentary structures associated with primary aeolian depositional processes as well as experimental and theoretical investigations of aeolian deposition (Hunter 1977).

Studies of dune processes

Studies of the physics of grain movement by the wind provide the basis of our knowledge of aeolian processes and sand transport rates. Many investigations are still strongly influenced by the seminal work of Bagnold (1941). In recent years debate on the nature of aeolian processes on other terrestrial planets, notably Mars and Venus, has led to a re-examination of the physics of grain movement and the nature of aeolian saltation. Consideration has been given to the effects of grain density and mineralogy and grain shape on sand transport rates.

Investigations of dune processes have concentrated on documenting the movement and dynamics of dunes in terms of airflow and sediment transport. Studies of the movement of dunes, especially those of barchan type, have had a long history. In recent years, there have been important advances in the knowledge of dune processes through careful study of winds and sand movements on individual dunes (e.g. Tsoar 1978; Havholm and Kocurek 1988; Lancaster 1989a; Livingstone 1989). These studies have given rise to a new understanding of the factors which influence dune dynamics and morphology of barchan and linear dunes, and demonstrate the importance of secondary airflow in controlling dune morphology.

As observed by Warren (1969), investigations of dunes are geographically clustered. In part there is also a temporal pattern. For example, field investigations of the Saharan sand seas, mostly by the French, peaked in the 1950s and 1960s. Since then there have been few field studies, but many investigations using remote sensing imagery. In recent years the best-studied sand seas and dune fields have been in Australia (especially the Simpson-Strzelecki), the Namib and Kalahari in southern Africa, the Thar Desert of India, and the deserts of the USA. Many Middle Eastern and Asian sand seas are almost unknown to modern investigators, except through remote sensing investigations.

CONTEMPORARY PARADIGMS FOR DUNE STUDIES

Today, the combination of the regional perspective of remote sensing data with detailed field studies has resulted in a new set of paradigms that emphasize studies of dune dynamics and processes on different temporal and spatial scales, in contrast to the descriptive approach employed by many earlier workers. Valuable perspectives are also coming from the realization

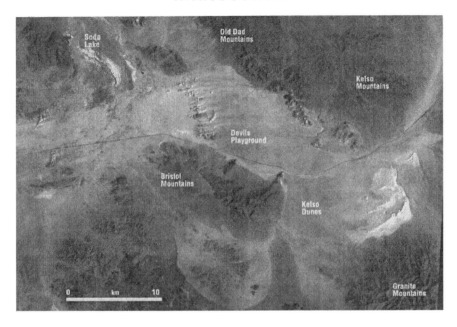

Plate 4 The Kelso Dunes aeolian sediment transport system. Sediment is transported from fluvial source areas to the west at the terminal fan-delta of the Mojave River toward the depositional sink, Kelso Dunes.

that Quaternary climatic and sea level changes have had a major effect on the accumulation of many sand seas.

Dune-forming processes operate at three main spatial and temporal scales (Warren and Knott 1983) which correspond approximately to the steady, graded and cyclic time scales of Schumm and Lichty (1965). Processes operating at the steady scale, which involves very short or even instantaneous amounts of time and small areas, include the formation of wind ripples. The graded scale involves periods of 10^{-1} to 10^2 years, and particularly concerns the dynamics and morphology of dunes, which tend towards an actual or partial equilibrium with respect to rates and directions of sand movements generated by surface winds. Form–flow interactions and feedback processes are important at this scale, which is probably the most important for determining the morphology and dynamics of dunes. The cyclic time scale involves periods of 10^3–10^6 years and a spatial scale corresponding to that of sand seas and their regional physiographic and tectonic setting. Processes at this scale involve those responsible for the accumulation of sand seas as sedimentary bodies.

At all scales of investigation, aeolian processes operate within geomorphic systems that form an interrelated set of processes and landforms in which sediment is transported by the wind from source areas to depositional sinks via transport pathways (Plate 4). Sources and sinks for sediment are linked by a cascade of energy and materials which can be viewed in terms of sediment inputs and outputs, transfers and storages. Study of aeolian processes within a systems framework permits the assessment of responses to external changes on differing spatial and temporal scales, as well as focusing attention on the links between system components.

This book is intended to give the reader an overview of modern concepts of dune forms, processes, and dynamics at the different spatial and temporal scales outlined above, beginning with sand transport by the wind, and ending with the development of sand seas, and the role of climatic changes. In doing this, I have drawn on my experience in sand seas in southern Africa and North America, as well as the work of many colleagues and co-workers.

2

SAND TRANSPORT BY THE WIND

INTRODUCTION

As will be demonstrated in subsequent chapters, spatial and temporal variations in sand transport rates play a major role in determining the dynamics and morphology of desert dunes. It is therefore necessary to understand the principles governing the entrainment and transport of sand by the wind before examining dune processes and dynamics.

Transport of sediment by the wind involves interactions between the wind and the ground surface. Understanding the operation of these processes requires a knowledge of relevant surface characteristics (e.g. sediment texture, vegetation cover, degree of cohesion and crusting) as well as the dynamics of airflow over the surface. There are three distinct modes of aeolian transport (Figure 2.1). These depend primarily on the grain size of the available sediment (Bagnold 1941). Very small particles (< 60–70 μm) are transported in suspension and kept aloft for relatively long distances by turbulent eddies in the wind. Particles of this size range play a minor or

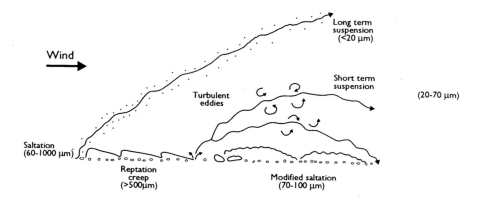

Figure 2.1 Modes of aeolian sediment transport (after Pye 1987).

insignificant role in dune dynamics, although post-depositional modification of dune sediments by infiltration of dust is important in many areas. Larger particles (approximately 60–500 µm, or sand size) move downwind by saltation (a series of short-distance jumps). The impact of saltating particles with the surface may cause short distance movement of adjacent grains (reptation). Larger (> 500 µm) or less exposed particles may be pushed or rolled along the surface by the impact of saltating grains in surface creep.

The dominant process of sand transport by the wind is therefore saltation, in which a cloud of grains is transported close to the surface by the shearing action of the wind. Four distinct sub-processes can be identified in the sand transport system (McEwan and Willetts 1993): aerodynamic entrainment, the trajectory of the wind-driven sand grains, the collision between these grains and the bed, and the modification of the wind by the saltation cloud.

THE SURFACE WIND

Naturally occurring airflow is almost always turbulent and consists of eddies of different sizes that move with different speeds and directions. In these conditions the flow can only be characterized by time-averaged parameters. Turbulent eddies transfer momentum from one 'layer' to another such that each 'layer' has a different average velocity and direction. This process is known as 'turbulent mixing' and the description of wind profiles in this way is known as the mixing length model (Prandtl 1935). As a result of frictional effects, the wind speed near the surface is retarded. If the surface is composed of very fine particles (< 80 µm) the surface is considered to be aerodynamically smooth. A very thin laminar flow sub-layer, usually less than 1 mm thick, develops adjacent to the bed even for flows in which most of the boundary layer is turbulent. By contrast, when the particles forming the surface or other roughness elements are relatively large (> 80 µm) the surface is regarded as aerodynamically rough. The laminar sub-layer ceases to exist and is replaced by a viscous sub-layer for which the velocity profile is not well understood (Middleton and Southard 1984). Under conditions of neutral atmospheric stability (defined as a condition in which the lapse rate equals the dry adiabatic lapse rate), the velocity profile above the viscous sub-layer for aerodynamically rough surfaces is characterized by the Prandtl–von Karman equation:

$$\frac{u}{u_*} = \frac{1}{k} ln \frac{z}{z_o}$$

(2.1)

where u is the velocity at height z, z_o is the aerodynamic roughness length of the surface, u_* is the shear velocity and k is von Karman's constant (~ 0.4). In these conditions, the wind profile plots as a straight line on semi-

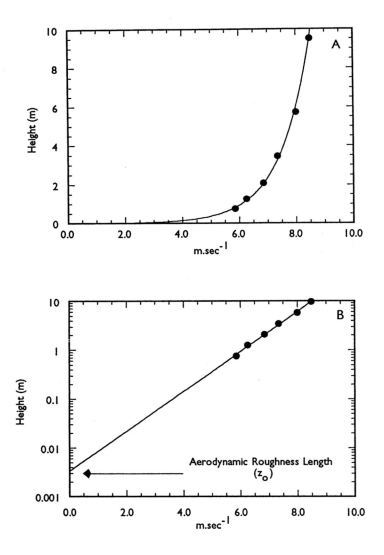

Figure 2.2 Boundary-layer wind profiles over the surface of an alluvial fan in Death Valley. A: linear scale; B: logarithmic scale for height, *y* intercept is an estimate of the aerodynamic roughness length.

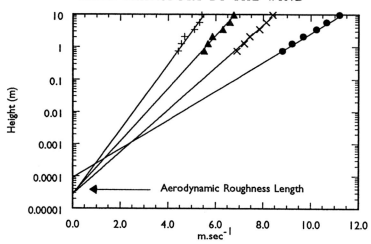

Figure 2.3 Changes in boundary-layer wind profiles over the surface of an alluvial fan in Death Valley with different wind speeds (measured at the highest anemometer). Note that the roughness length (z_o) remains the same although wind speed at a given height increases.

logarithmic axes with the y intercept representing the aerodynamic roughness length (z_o) (Figure 2.2). No matter how strongly the wind blows all the velocity profiles tend to converge to the same intercept or roughness length (Figure 2.3). Thus u_* increases with increasing wind speed at a given height (z) above the surface.

For sand surfaces, the aerodynamic roughness length (z_o) is approximately 1/30 the mean particle diameter. For rougher surfaces, the roughness length also varies with the shape and distance between individual particles or other roughness elements so that z_o increases to a maximum of ~ 1/8 particle diameter when the roughness elements are spaced ~ 2 times their diameter (Greeley and Iversen 1985). Representative values of aerodynamic roughness for unvegetated desert surfaces are shown in Table 2.1. The exact nature of the relationship between z_o and surface roughness and particle size is however poorly understood (Lancaster *et al.* 1991; Blumberg and Greeley 1993).

Table 2.1 Typical values of aerodynamic roughness (z_o) for desert surfaces.

Surface type	Aerodynamic roughness (m)
Playas	0.00008–0.00013
Alluvial fans	0.00076–0.00310
Lava flows	0.00132–0.07351

Source: From Blumberg and Greeley 1993.

13

Where the surface is covered by tall vegetation or high densities of other large roughness elements (e.g. on lava flows), the point at which the wind speed is zero is no longer the ground surface, but is displaced upwards from the surface to a new reference plane which is a function of the height, density, porosity and flexibility of the roughness elements (Oke 1978; Wolfe 1993). The upward displacement is termed the zero plane displacement height (d), which is thought to represent the mean level of momentum absorption. The wind profile equation then becomes:

$$\frac{u}{u_*} = \frac{1}{k} ln \frac{z - d}{z_o}$$

(2.2)

However, the significance of the zero plane displacement height has been questioned for sparsely vegetated desert surfaces (Wolfe and Nickling 1993).

The shear velocity (u_*) is proportional to the slope of the wind velocity profile when plotted on a logarithmic height scale (Figure 2.2) and is related to the shear stress (τ) at the bed and the air density (ρ_a) by

$$u_* = \sqrt{\frac{\tau}{\rho_a}}$$

(2.3)

These relations represent the ideal case of a wind blowing across a horizontal homogeneous surface in conditions in which the atmospheric stability is neutral (lapse rate equal to the dry adiabatic lapse rate: DALR). In practice, however, atmospheric conditions in desert regions are frequently unstable (lapse rate > DALR) as a result of surface heating. This results in enhanced convection and vertical mixing so that the velocity gradient changes little with height above the surface (Figure 2.4). As a result, the shear stress in this region will decrease, but the near-surface shear stress will increase because higher wind speeds occur nearer the surface (Frank and Kocurek 1994a). In stable conditions, vertical mixing and convection are reduced, so increasing shear velocity away from the surface. These effects are significant when wind speeds (at a height of 5 m) are less than 9.5 m.sec^{-1} in unstable and 13.5 m.sec^{-1} in stable conditions (Frank and Kocurek 1994).

Another important effect on wind profiles is that of sloping surfaces. As discussed in detail in Chapter 5, streamline convergence and flow acceleration on the windward slopes of dunes result in wind profiles that are no longer log-linear, so that mixing length models for wind profiles cannot be used in these situations.

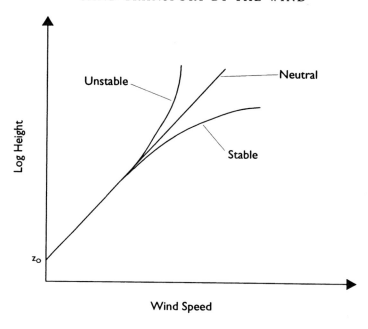

Figure 2.4 Effects of atmospheric stability on the wind profile (after Oke 1978).

ENTRAINMENT OF SAND BY THE WIND

The threshold of motion

Sand grains will be moved by the wind when the fluid forces (lift, drag, moment) exceed the effects of the weight of the particle and cohesion between adjacent particles (Figure 2.5). The drag and lift forces as well as the resultant moment are caused by the fluid flow around and over the exposed particles. The lift force results from the decreased fluid static pressure at the top of the grain as well as the steep velocity gradient near the grain surface. The weight and cohesive forces are related to physical properties of the surface particles including their size, density, mineralogy, shape, packing, moisture content, and the presence or absence of bonding agents such as soluble salts.

As drag and lift on the particle increase, there is a critical value of wind shear velocity when grain movement is initiated. This is the fluid threshold shear velocity or u_{*t} (Bagnold 1941):

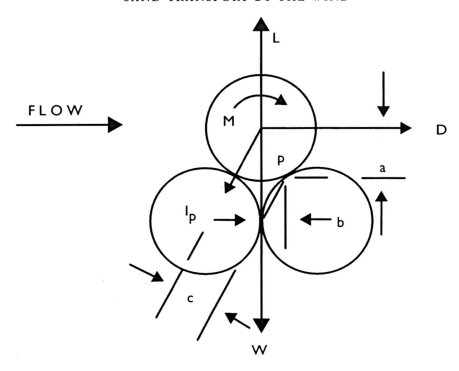

Figure 2.5 Schematic view of forces on a spherical particle at rest: D = aerodynamic drag, L = lift, M = moment, I_p = interparticle force, and W = weight. Moment arms about a point p given by a, b, c (after Greeley and Iversen 1985).

$$u_{*t} = A\sqrt{\frac{(\rho_p - \rho_a)}{\rho_a}g},$$

(2.4)

where A is an empirical coefficient dependent on grain characteristics (approximately 0.1 for sand-sized particles), and D is particle diameter.

During the downwind saltation of grains their velocity and momentum increases before they fall back to the surface. On striking the surface, the moving particles may bounce off other grains and become re-entrained into the airstream or embedded in the surface. In both cases, momentum is transferred to the surface in the disturbance of one or more stationary grains. As a result of the impact of saltating grains, particles are ejected into the airstream at shear velocities lower than that required to move a stationary grain by direct fluid pressure. This new lower threshold required to move stationary grains after the initial movement of a few particles is the dynamic or impact threshold (Bagnold 1941). The dynamic or impact threshold for a given sediment follows the same relationship as the fluid threshold (Equation 2.4) but with a lower coefficient A of 0.08 (Figure 2.6).

Entrainment processes

Bagnold (1941) suggested that once the critical threshold shear velocity is reached, stationary surface grains begin to roll or slide along the surface by the direct pressure of the wind. Once particles begin to gain speed they start to bounce off the surface into the airstream and initiate saltation. More recent work (e.g. Iversen and White 1982) indicates that particles moving into saltation do not usually roll or slide along the surface prior to upward movement into the airstream. Initial movement into saltation is caused by instantaneous air pressure differences near the surface which act as lift forces. As wind velocity is increased, particles begin to vibrate with increasing intensity and at some critical point leave the surface instantaneously (Lyles and Krauss 1971). The number of grains entrained aerodynamically is given by:

$$N_a = \alpha \, (\tau_a - \tau_t) \tag{2.5}$$

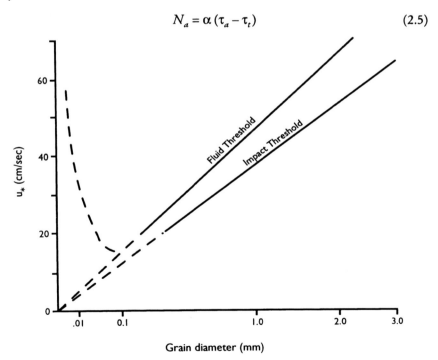

Figure 2.6 Relations between fluid and impact threshold shear velocity and particle size (after Bagnold 1941).

17

where N_a is the number of grains leaving the surface per unit area, α is a constant with units of m.s.kg^{-1}, τ_a is the mean fluid shear stress at the surface and τ_t is the threshold shear stress (Anderson and Haff 1988; McEwan et al. 1992).

Williams et al. (1990), Willetts et al. (1991) and Williams et al. (1994) show that fluid threshold is dependent on the turbulent structure of the boundary layer, and that initial disturbance is related spatially and temporally to semi-organized 'flurries' of activity (Lyles and Krauss 1971) associated with sweep and burst sequences well known in boundary layers, as well as to spatial inhomogeneity in the boundary layer. Willetts et al. (1991) suggest that it is the peak rather than the mean value of short-term shear stress that determines the level of aerodynamic entrainment. Grains dislodged by sweeps of movement then become the agents for further dislodgement by collision.

The threshold process is therefore complex, involving localized near-bed turbulent features that dislodge grains which act as 'seeds' for downwind ejections of further grains at shear velocities lower than that required to entrain them by direct fluid pressures and lift forces. The newly ejected particles move downwind and impact the surface, displacing an even larger number of stationary grains. Initiation of widespread grain motion therefore involves a cascading effect, such that the number of grains in motion increases exponentially (Nickling 1988). Interactions between the developing saltation cloud and the wind result in extraction of momentum from the near-surface wind and a reduction in the near-bed wind speed so that the saltating grains are accelerated less and impact with less energy. As a result, the number of grains dislodged at the bed decreases, and the transport rate approaches a dynamic equilibrium value termed 'steady state saltation' (Anderson and Haff 1988) or 'equilibrium saltation' (Owen 1964). Equilibrium appears to be reached very rapidly, with a characteristic time period of 1–2 seconds (Figure 2.7). More recent work (Willetts et al. 1991; McEwan and Willetts 1993) suggests that the attainment of equilibrium saltation is a two-stage process in which the wind and the saltation cloud first reach an equilibrium on a time scale of 1–2 seconds, and then the wind profile adjusts to the increased surface roughness over a period of several tens of seconds. McEwan and Willetts (1993) regard the latter condition as the true equilibrium condition. This implies that sand transport rates in natural conditions are unlikely to be in complete equilibrium with flow conditions. Measurements of sediment flux are therefore the average of a series of temporally adjusting flows.

Effects of surface conditions on entrainment

Although fluid threshold can be closely defined for a uniform sediment size greater than approximately 100 μm, this is not the case for most natural sediments because they usually contain a range of grain sizes and shapes

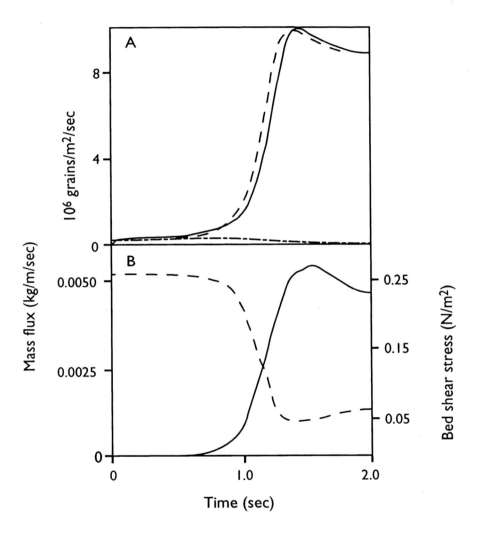

Figure 2.7 Computer simulations of saltation for 25 µm grains. A: numbers of grains impacting (dashed line), being ejected (rebound and splash, solid line), and being entrained (dot-dashed) as a function of time. Note that equilibrium is reached in a very short time (~ 2 seconds). B: corresponding history of total mass flux (solid) and shear stress at the bed (dashed) (after Anderson and Haff 1988).

which vary in grain density and packing. As a result, fluid and dynamic thresholds should be considered as threshold ranges that are a function of the size, shape, sorting and packing of the surface sediments.

Grain size effects

The relationship between threshold shear velocity and particle diameter based on Equation 2.4 (Figure 2.6) shows that, for grains greater than 80 μm diameter, u_{*t} increases with the square root of grain diameter. In this case, the grains protrude into the airflow and are therefore aerodynamically 'rough' so that the drag acts directly on the grains. Smaller grains lie within the laminar sub-layer. The drag is distributed more evenly over the surface which is aerodynamically 'smooth', and the value of the threshold parameter A rises rapidly. As a result u_{*t} is no longer proportional to the square root of grain diameter but is dependent on the value of A. Wind tunnel studies (e.g. Bagnold 1941; Chepil 1945a, b) confirm that very high wind speeds are required to entrain fine grained materials. However, Greeley et al. (1976) and Iversen and White (1982) have questioned the assumption that the coefficient A in the Bagnold threshold model is a unique function of particle size. They indicate that A is nearly constant for larger particles but increases rapidly for small particles, suggesting that it is not solely a function of particle diameter for grains less than about 80 μm in diameter. For very small grains inter particle forces such as electrostatic charges and moisture films appear to be important in determining the threshold for entrainment.

Local bed slope

Sand transport on dunes almost always occurs on sloping surfaces. The local bed slope may influence threshold shear velocity through the effects of gravity acting via the angle of internal friction of the sediment. Howard (1977) produced a theoretical expression for the effect of local slope on threshold shear velocity:

$$u_{*t} = F^2 D [(\tan^2 \alpha \cos^2 \theta - \sin^2 \chi \sin^2 \theta)^{1/2} - \cos \chi \sin \theta] \qquad (2.6)$$

where $F = B(g \, \rho_s/\rho)^{1/2}$, B is a dimensionless constant with a value of 0.31, D is the grain diameter, α is the angle of internal friction of the sediment, θ is the local slope and χ is the angle between the local wind direction and the direction normal to the maximum local slope.

Hardisty and Whitehouse (1988a, b) used a portable wind tunnel on Saharan sand dunes to investigate the effects of slope on threshold shear velocity. They found that threshold increases slightly with slope for positive (upward) sloping surfaces, and decreases significantly for negative (downward) slopes (Figure 2.8).

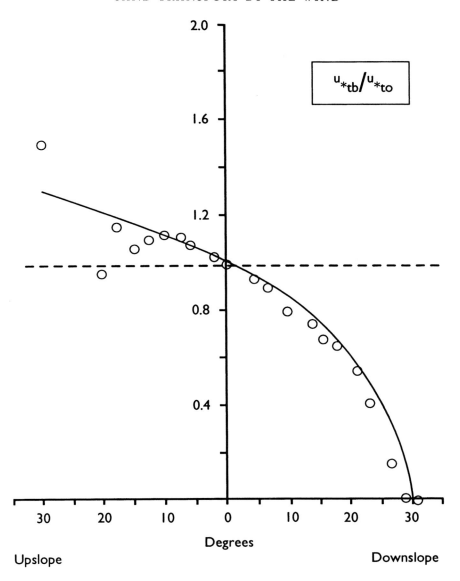

Figure 2.8 Relationship between threshold velocity (u_{cr}) and bed slope. Circles are values from field experiments, solid line is theoretical relationship determined by Dyer (1986). u_{*tb}, u_{*to} the ratio between threshold velocity on a sloping bed and a flat bed (equal to unity on a flat bed) (after Hardisty and Whitehouse 1988a). Reprinted with permission from *Nature*.

Moisture content

Surface moisture content is an extremely important variable controlling both the entrainment and flux of sediment by the wind. Capillary forces developed at inter-particle contacts increase the cohesion and thus the threshold shear velocity. Gravimetric moisture contents of approximately 0.6 per cent can more than double the threshold velocity of medium sized sands. Above approximately 5 per cent gravimetric moisture content, sand-sized material is inherently resistant to entrainment by most natural winds (Belly 1964).

The capillary force moment for grains with either open or closed packing arrangements can be incorporated into Bagnold's (1941) threshold equation to provide a general expression for the threshold velocity of a moist surface (McKenna–Neuman and Nickling 1989):

$$u_{*tw} = A \left[(\rho_p - \rho_a) \right]^{1/2} \left[6 \sin 2\alpha / (\pi D^3 (\rho_p - \rho_a) g \sin \alpha) F_c + 1 \right]^{1/2} \quad (2.7)$$

where F_c is the capillary force moment, α is the angle of internal friction and D is grain diameter.

Bonding agents and surface crusts

Particles at the surface may be bound together by silt and clay, organic matter, or precipitated soluble salts to produce erosion-resistant structural units or more continuous surface crusts (Chepil and Woodruff 1963; Nickling 1984). Even low concentrations of soluble salts (especially sodium chloride) can significantly increase threshold velocity by the formation of cement-like bonds between individual particles (Nickling and Ecclestone 1981). Surface crusts formed by raindrop impact, algae and fungi, or precipitation of soluble salts, can also increase the threshold velocity (Gillette et al. 1982).

Surface roughness

The presence of vegetation, gravel lag deposits or other non-erodible roughness elements is important to both the entrainment and transport of sand by the wind. When the wind blows over a smooth unobstructed surface, the shear stress is imparted more or less uniformly across the entire surface, but when non-erodible roughness elements are present a proportion of the shear stress is absorbed by the roughness elements protecting the underlying erodible surface. The degree of protection is a function of their size, geometry, and spacing (Marshall 1971; Lyles et al. 1974; Gillette and Stockton 1989; Musick and Gillette 1990). Low densities of roughness elements (glass spheres and gravel particles) tend to reduce the threshold velocity of the surface and cause increased erosion around the elements because of the development and shedding of eddies (Logie 1982). By contrast, higher densities of roughness elements tend to increase the threshold velocity of the surface.

SALTATION PROCESSES

The process of sand transport by the wind can be viewed as a cloud of grains saltating along a sand bed, with grains regaining from the wind the momentum lost by rebounding from the bed. These high energy grains rebound from the bed in a process termed 'successive saltation' (Rumpel 1985). Impacts between the saltating grains and the bed transfer momentum to other grains, which move a short distance. The motion of these low-energy ejected grains is termed reptation (Ungar and Haff 1987). The number and velocity distribution of grains ejected in this way is described statistically by the 'splash function' (Ungar and Haff 1987; Werner 1988). The saltating grains rebound with ~ 50–60 per cent of their original velocity (Willetts and Rice 1986; Anderson and Haff 1988). However, the ejected grains reach a maximum of only ~ 10 per cent of the speed of the impacting grain. In addition, rearrangement of the bed by saltation impacts leads to movement of grains that remain in contact with the bed at all times (surface creep) (Willetts et al. 1991).

Saltation trajectories

Sand particles moving in saltation are characterized by trajectories with an initial steep vertical ascent followed by a parabolic path such that they return to the bed with relatively small impact angles (Figure 2.9). There are four forces that determine the trajectory: (1) gravity, (2) aerodynamic drag, (3) the Magnus effect, and (4) aerodynamic lift. For an idealized case assuming that the initial take-off speed is in the same order as the shear velocity (u_*) the trajectory height is 0.81 u_*^2/g and the length 10.3 u_*^2/g (Owen 1964). Lift-off angles of 75° to 90° were measured by Bagnold (1941) and Chepil (1945b), but White and Schultz (1977) determined the average ejection angle as ~ 50° for particles 586 μm in diameter for a shear velocity of 40 cm.sec^{-1}. White and Schultz (1977) also observed that particle trajectories were higher than those predicted by theoretical equations of motion. This occurs because of a lift force that is generated by the spinning of the grain as it moves through the air (the Magnus effect) which is related to the rate of spinning (350 to 400 rps) and the steep velocity gradient adjacent to the particle surface (White 1982). When the Magnus effect is included in the equations of motion, theoretical trajectories are in much better agreement with those observed in wind tunnel experiments (White and Schultz 1977).

Because the near-surface wind velocity gradient is steep, the higher the saltating particles rise from the surface, the greater the velocity at which they

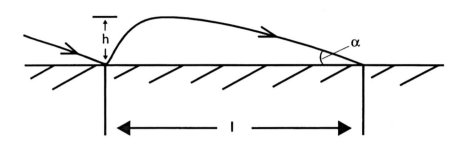

Figure 2.9 Schematic view of a typical saltation trajectory. h = saltation height, α = impact angle, and l = saltation path length.

will be carried in the airstream. This also gives rise to longer saltation path lengths. Once particles have attained their maximum height they descend approximately linearly, to impact with the surface at incidence angles of 10° to 16° (Bagnold 1941), or from 4° to 28° with an average angle 13.9° (White and Schultz 1977). The impact angle of a saltating particle tends to decrease with an increase in wind velocity and particle size (Sorensen 1985). Grain shape has a pronounced effect on the saltation path and the nature of the bed collisions. Platy sands tend to saltate in longer, lower trajectories compared to more spherical particles (Willetts 1983) and collision is a more efficient mechanism for maintaining saltation in compact quartz particles than for platy sands (Willetts and Rice 1986).

Modification of the surface wind by saltating grains

When the wind shear velocity is great enough to move sand particles the near-surface wind velocity profile is altered because momentum is extracted from the wind near the surface by the saltating sand grains. This decelerates the near-surface wind. In addition, the fluid shear stress is modified by the grains so that it is no longer constant with height (the grain-borne shear stress of Owen (1964)). The velocity profile in the saltation layer is therefore modified from the logarithmic one found above this layer, so that the profiles exhibit a distinct 'kink'. Bagnold (1941) showed that velocity distributions with height remain as a straight line but tend to converge to a focus (z_o') at a point 0.2–0.4 cm above the surface (Figure 2.10). Below this the wind profile is non-logarithmic, and wind speed in this zone may actually decrease

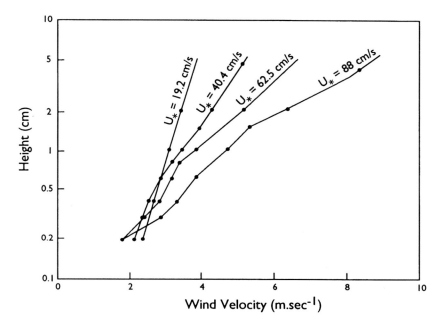

Figure 2.10 Change in near-bed wind profiles and roughness length under conditions of sand transport. The straight profile with a u_* of 19.2 cm/sec is at fluid threshold. All the other profiles are for winds above threshold and display a 'kink' at 0.5 to 1 cm above the surface (modified from Bagnold 1941).

as u_* increases (Bagnold 1941). The focus (z_o') may represent the mean saltation height of uniformly sized grains (Bagnold 1941), but Anderson and Haff (1988) have shown that there is a range of saltation trajectories. Owen (1964) believed that the saltation layer acted to increase aerodynamic roughness by an amount related to the mean vertical lift-off velocity of the grains, so that the apparent bed roughness (z_o') is dependent on u_* during sediment transport:

$$z_o' = \alpha \frac{u_*^2}{2g}$$

(2.8)

where $\alpha = 0.02$ when determined from wind tunnel experiments.

Detailed analyses of near bed wind profiles by Gerety (1985) also cast doubt on Bagnold's concept of a focal point, suggesting instead a focal zone at 2–3 cm, below which the grains significantly alter the flow, and above which the grain concentrations are too low to be significant. Numerical models of saltation (McEwan and Willetts 1991) discussed in McEwan (1993)

25

show however that the kink in the profiles identified by Bagnold is a real physical feature that is caused by a maximum in the extraction of momentum from the wind by saltating grains (Figure 2.11).

The saltation layer

Most sediment in the cloud of saltating and reptating grains is transported close to the ground, with an exponential decline in sediment and mass flux concentration with height (Bagnold 1941; Sharp 1964; Williams 1964; Nickling 1983; Anderson and Hallet 1986) (Figure 2.12). Over sand surfaces most grains travel within the lower 1–2 cm (Bagnold 1941, Williams 1964),

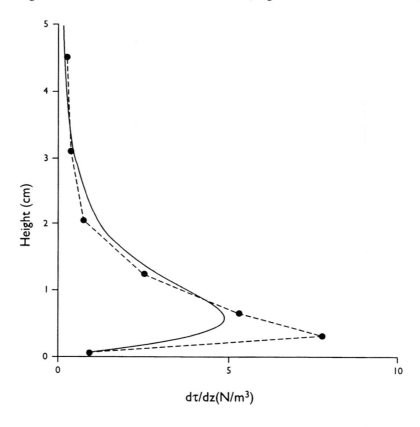

Figure 2.11 Numerical model for the effects of saltation on the wind profile. $d\tau/dz$ (derived from the hyperbolic tangent stress profile, solid line) is the force per unit volume exerted by the grains on the wind and shows a maximum at 5.77 mm. This is the cause of the kink in the wind profiles in Figure 2.10. The broken line is the stress profile predicted by the numerical model of McEwan and Willetts (1991) (after McEwan 1993).

26

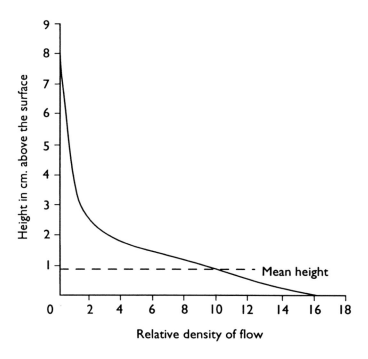

Figure 2.12 Relationship between mass flux of material in transport and height above the bed (after Bagnold 1941).

but reach greater heights over hard gravel-covered surfaces. Sharp (1964) found that 50 per cent by weight of sediment transported across an alluvial surface in the Coachella Valley travelled within 13 cm of the ground, and 90 per cent below 60 cm. Williams (1964) and Gerety and Slingerland (1983) also observed a decline in average grain size with height. However, the size of sediment transported at a given height tends to increase with u_*.

SAND TRANSPORT EQUATIONS

The seminal work of Bagnold (1941) relating the quantity of sand transported as a function of the shear stress exerted by wind forms the basic theoretical background and supporting empirical evidence for almost all research on aeolian sand transport rates. Using theoretical considerations in conjunction with field and wind tunnel observations Bagnold found that the sediment transport mass flux in saltation and creep (q) could be defined by:

$$q = \sqrt{C \frac{dp}{D} \frac{\rho}{g} u_*^3}$$

$$(2.9)$$

where (dp/D) is the ratio of the mean size of a given sand to that of a 'standard' 0.25 mm sand. The coefficient C is a sorting coefficient with values of 1.5 for nearly uniform sand, 1.8 for naturally graded sand, 2.8 for poorly sorted sand, and 3.5 for a pebbly surface. For a given value of u_*, the sediment flux, q, increases from a minimum for nearly uniform sand to somewhat higher values for more poorly sorted sands and reaches a maximum value over a pebble surface. The effect of impacts from larger saltating grains is probably greater for poorly sorted sands. Sand grains saltate more effectively and further across hard and pebbly surfaces because less momentum is extracted by impacts with the bed.

Following Bagnold's (1941) work, other investigators have developed both theoretical and empirical equations to describe the transport of sediment by wind (for a review see Sarre 1989). These equations, although frequently derived in different ways and from different points of view, are similar in general form to that initially proposed by Bagnold (1941).

A fundamental problem with the original Bagnold equation is that it does not include a threshold term and thus predicts sediment transport at shear velocities below that required to initiate particle movement, although Bagnold (1956) did modify his equation to include such a term. Kawamura (1951) proposed a somewhat different equation that included a threshold shear velocity (u_{*t}) term:

$$q = K \frac{\rho_a}{g} (u_* - u_{*t}) (u_* - u_{*t})^2$$

$$(2.10)$$

where K is an empirical coefficient which is a function of the textural characteristics of the sediment ($K = 2.78$ for moderately well-sorted sand with a mean diameter of 0.25 mm). A threshold shear velocity term was also included in the transport equation derived by Lettau and Lettau (1978):

$$q = C \frac{dp}{D} \frac{\rho}{g} u*^2 (u_* - u_{*t})$$

$$(2.11)$$

where the exponent n has values ranging from 0.5 to 0.75. A universal transport equation was proposed by (White 1979):

$$q = 2.61 u_*^2 (1 - u_{*t}/u_*) (1 + u_{*t}^2/u_*^2) \rho_a/g$$

$$(2.12)$$

This expression provides good estimates of actual sediment flux as demonstrated by both wind tunnel and field experiments (Greeley and Iversen, 1985; Greeley et al. in press).

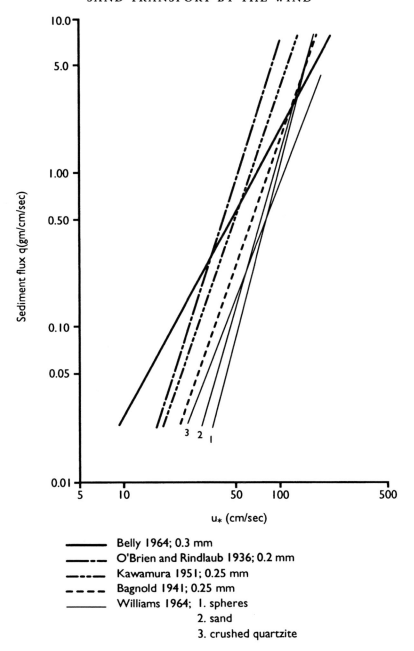

Figure 2.13 Comparison between different flux equations for sand-sized sediment (after Lancaster and Nickling 1994).

Comparison of the sediment flux predicted by different transport equations with increasing shear velocity for similar particle sizes (Figure 2.13) shows that despite the rather similar form of the equations and similar mean particle diameters used in the calculations, the equations give significantly different results. However, accurate data on actual sediment fluxes are difficult to obtain, due to less than perfect sand trap efficiency and the non-stationary nature of the bed in wind tunnel experiments (Rasmussen and Mikkelsen 1991; Nickling and McKenna-Neuman 1995).

Data on wind shear velocity are rarely available outside of specific field or wind tunnel experiments, yet there is a clear need to be able to estimate sand transport rates from conventional meteorological measurements of wind velocity at a single height in order to assess sediment transport on a regional scale. Bagnold (1941) was aware of this problem and modified his flux equation (Equation 2.10) accordingly to express transport rate as a function of wind speed (v) at 1 metre above the surface:

$$q = 1.5.10^{-9} (v - V_t)^3 \qquad (2.13)$$

assuming a constant roughness length. Bagnold (1953) later modified this expression for general use to:

$$q = (1.0.10^{-4}/(\log 100\ z)^3).t.(v - 16) \qquad (2.14)$$

where q is the flux in tonnes/metre width, z is the wind measurement height, t is the number of hours a wind with a velocity v km.hr^{-1} blows and 16 km.hr^{-1} is a threshold velocity (V_t). This formula was used to assess regional sand transport rates in the Namib and Kalahari by Lancaster (1981a, 1985a). Hsu (1971) derived a different empirical formula for sand transport in terms of wind velocity at 10 m height (the World Meteorological Standard Height):

$$q = 1.16.10^{-4}\ v^3 \qquad (2.15)$$

assuming a value for u_* of 4.0 m.sec^{-1}.

In addition to the very sparse distribution of meteorological observations of wind speeds in most arid regions, the major problem with using wind velocity measurements to estimate sand transport rates on a regional scale is that u_* varies with the aerodynamic roughness of the surface, which changes from place to place. Use of orbital radar data to assess aerodynamic roughness may assist in providing better data for regional sand transport estimates (Greeley et al. 1991). Further, unless wind profiles are corrected for atmospheric stability effects, calculations of sediment fluxes may be in error by as much as 15 times (Frank and Kocurek 1994a).

All equations for sand transport should be regarded as providing maximum potential rates at which the capacity of the wind for transport is saturated (Wilson 1971). Actual rates are limited by sediment availability, and are frequently less than the maximum, or under saturated.

CONTROLS ON SAND TRANSPORT RATES

Textural effects

Most sediment transport equations contain various empirically derived coefficients which are a function of the grain size and sorting characteristics of the sediments. Differences in the sediment flux predicted by these equations may be related to undocumented textural differences, especially the grain shape of the test sands used in the experiments (Williams 1964; Willetts et al. 1982). Sediment flux (q) varies as a power function of shear velocity (u_*) but with exponents considerably different from the value of 3 suggested by Bagnold (1941) and others. The value of the exponent and thus the flux of sediment increases with the sphericity of the sediment from 2.76 for crushed quartzite to 3.42 for natural sand, and 4.10 for glass spheres (Williams 1964). Williams (1964) also showed that at low shear velocities (u_* < 75 cm.sec^{-1}) there is a tendency for sediment transport rate to increase as particle shape becomes more irregular.

Effects of moisture content

Although dune sand in many desert areas is normally dry, damp sand may exist in the subsurface in semi-arid areas (Hyde and Wasson 1983). Moisture affects sand transport rates through the higher threshold velocity required to entrain damp sand. As wind shear velocities increase, the effects become less (Figure 2.14), although transport rates are still lower by as much as 25 per cent for a wind shear stress of 1 m.sec^{-1} (Hotta et al. 1984; Sherman 1990). Comparisons of calculations of transport rates using the Bagnold formula and dune migration data on the Oregon Coast indicate that in this area actual transport rates are reduced by as much as one-third by wet conditions (Hunter et al. 1983). However, the exact nature of the relationships between moisture content and transport rates are poorly known, and the role of evaporation in the process is uncertain (Hyde and Wasson 1983; Sherman 1990).

Effects of vegetation

Dunes and aeolian depositional features that are vegetated to varying degrees occur in most deserts, including even the hyper-arid Namib. The effects of vegetation on sand transport rates are however poorly known. Empirical studies of the effects of vegetation on sand transport by Ash and Wasson

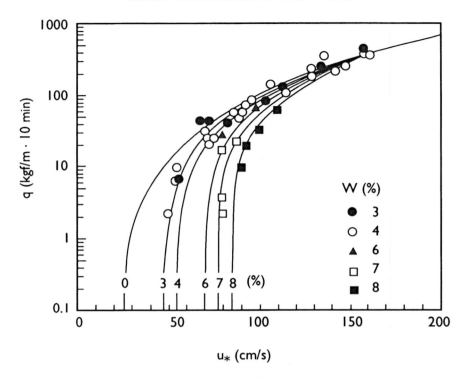

Figure 2.14 Relation between mass flux and moisture content (*W*)
(after Hotta *et al.* 1984).

(1983) and Wasson and Nanninga (1986) indicate that sand transport can take place even when vegetation cover is as much as 45 per cent.

A more rigorous approach involves assessment of the partitioning of wind shear stress between plants and the sand surface. In the absence of large roughness elements, the shear stress at the surface can be determined from the shear velocity (u_*) using the log-linear wind profile (Equation 2.1) by

$$\tau_s = \rho \, u_*^2 \qquad (2.16)$$

where τ_s is the shear stress at the surface. In the case of surfaces covered with large roughness elements, the logarithmic wind profile does not extend down between the roughness elements, so that computation of the surface shear stress from the wind profile is impossible. In these situations the total force (*F*) imparted to the surface is equal to the sum of the forces on the roughness elements F_r and the intervening surface F_g (Schlicting 1936) so that:

$$F = F_r + F_g \qquad (2.17)$$

Stockton and Gillette (1990) showed that the total shear stress on a surface covered with roughness elements can be partitioned between the elements and the intervening surface by:

$$\tau = F_r/A_s + (F_g/A_g)(A_g/A_s) \\ = F_r/A_s + \tau_g(A_g/A_s) \qquad (2.18)$$

where F_r is the force exerted on the roughness elements, A_g is the ground area not covered by roughness elements and τ_g is the shear stress on the intervening ground surface. The fraction of the total force going to the erodible surface (shear stress partitioning) is:

$$1 - [F_r/(A_s\,\tau)] = (\tau_g/\tau)(A_g/A_s) \qquad (2.19)$$

which can be rewritten as a shear stress ratio defined by

$$1 - [F_r/(A_s\,\tau)] = R^2(A_g/A_s) \qquad (2.20)$$

where R is the threshold velocity ratio (u_{*t}/u_{*tv}) (Musick and Gillette 1990). The two parameters expressing the stress partitioning in this equation are R, the ratio of threshold shear velocity without vegetation to the threshold shear velocity with vegetation, and A_g/A_s, a geometric factor describing the proportion of surface area covered by the erodible sediment. The shear stress ratio can be shown to be a function of the density and size of the vegetation cover as expressed by the lateral cover (L_c) (Figure 2.15).

Effects of roughness elements

For an unvegetated surface with non-erodible roughness elements, e.g. desert pavements, the relation between transport rates and the size, density, and spacing of roughness elements is influenced strongly by the effect of roughness element geometry on the threshold for transport. Greeley *et al.* (1974) summarized this as:

$$u_{*t}^2 = 0.139\,\rho g D / \rho(1 + 0.776\,(ln(1 + D/z_o))^2) \qquad (2.21)$$

where D = the diameter of the grains in transport.

Transport rates may therefore be larger or smaller on surfaces of different roughness, depending on the relative values of u_*, u_{*t}, and the ratio between particle diameter and z_o (Figures 2.16 and 2.17). There is a sharp increase in threshold for particles between 100 and 200 μm (Blumberg and Greeley 1993). Greeley and Iversen (1987) suggested that transport takes place only on smoother surfaces at low wind speeds; at intermediate speeds transport occurs on both surfaces, with greater transport on the smooth surface.

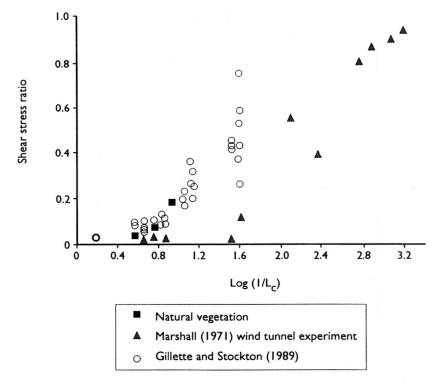

Figure 2.15 Relation between shear stress ratio and lateral cover of roughness elements (after Musick and Gillette 1990).

However, at high wind speeds, transport rates are greater on the rough surface.

Slope effects

Most dune surfaces are not flat, so the effect of slope on transport rates is potentially very important. Bagnold (1956) showed that the transport rate on an inclined surface, q_i, is proportional to:

$$q_i = q / (\cos \theta (\tan \alpha + \tan \theta)) \qquad (2.22)$$

where q is the transport rate on a flat surface, θ is the slope angle, and α is the angle of internal friction of the sand. Field experiments by Hardisty and Whitehouse (1988a) show that actual transport rates deviate from this relationship (Figure 2.18), suggesting an increased role of surface slope that

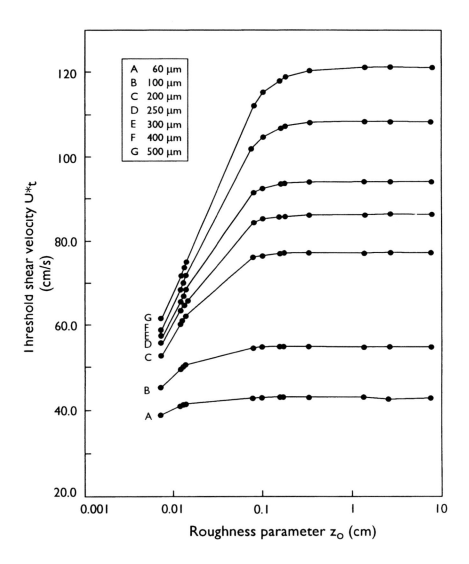

Figure 2.16 Relations between threshold shear velocity and aerodynamic roughness (after Blumberg and Greeley 1993).

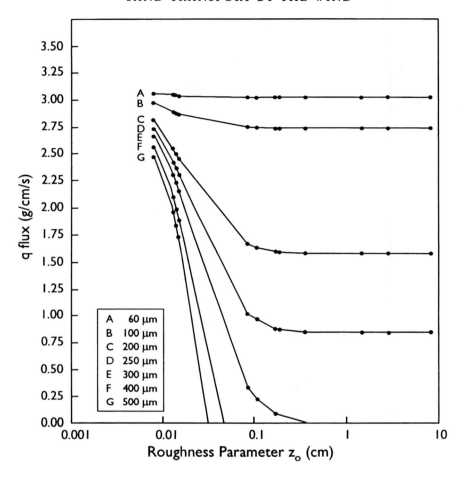

Figure 2.17 Relations between mass flux and aerodynamic roughness (after Blumberg and Greeley 1993).

significantly enhances sand transport on downward sloping surfaces, and retards it on upward sloping surfaces, due to the effects of gravity on grain collisions and the reptation process (Willetts and Rice 1988).

WIND RIPPLES

Wind ripples (Plate 5) are ubiquitous on all sand surfaces except those undergoing very rapid deposition. They are the initial response of sand surfaces to sand transport by the wind and form because flat sand surfaces

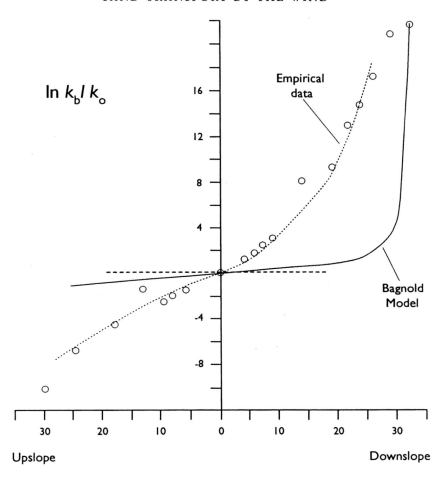

Figure 2.18 Relation between mass flux and slope (after Hardisty and Whitehouse 1988). Reprinted with permission from *Nature*.

over which transport by saltation and reptation takes place are dynamically unstable (Bagnold 1941). The formation and movement of wind ripples are therefore closely linked to the processes of saltation and reptation.

Wind ripples trend perpendicular to the sand transporting winds, although Howard (1977) has emphasized the effects of slope on ripple orientation. Because they can be reformed within minutes, wind ripples provide an almost instantaneous indication of local sediment transport and wind directions. Typical wind ripples have a wavelength of 50–200 mm and an

(a)

(b)

amplitude of 0.005–0.010 m (Bagnold 1941; Sharp 1963). Much longer wavelength (0.5–2 m) ripples with an amplitude of 0.1 m or more are composed of coarse sand or granules (1–4 mm median grain size) and are termed 'granule ripples' by Sharp (1963) and Fryberger *et al.* (1992) and 'megaripples' by Greeley and Iversen (1985). Ellwood *et al.*(1975) showed that granule ripples are not distinct forms, and form one end of a continuum of wind ripple dimensions (Figure 2.19). Ripple wavelength is a function of both particle size and sorting and wind speed so that ripples in coarse sands have a greater spacing than those in fine sands (Sharp 1963). For sands of a given size, ripple wavelength increases with wind shear stress (Seppälä and Linde 1978). Most wind ripples are asymmetric in cross-section with a slightly convex stoss slope at an angle of 8–10° and a lee slope that varies between 20 and 30° (Sharp 1963). In all cases the crest of the ripple is composed of grains that are coarse relative to the mean size of the surface sand.

Several models have been put forward to explain the formation and characteristic wavelength of ripples. Bagnold (1941) pointed to the close correspondence between *calculated* saltation path lengths and *observed* ripple wavelengths in wind tunnel experiments. Chance irregularities in 'flat' sand surfaces give rise to variations in the intensity of saltation impacts (Figure 2.20a) creating zones that would be preferentially eroded or protected. Grains from the zone of more intense impacts (A–B) would land downwind at a distance equal to the average saltation path length such that a zone of more intense saltation impacts would propagate downwind. In addition, variations in surface slope and saltation impact intensity cause variations in the reptation rate. Bagnold argued that interactions between the developing surface micro topography and the saltating and reptating grains would soon lead to a coincidence between the characteristic saltation path length and the ripple wavelength (Figure 2.20b). Wilson (1972) and Ellwood *et al.* (1975) extended Bagnold's hypothesis to show that even the wavelengths of granule ripples were related to the mean saltation path length.

Bagnold's concept of the formation of wind ripples was first challenged by Sharp (1963) who argued that grains in ripples are moved mostly by reptation. Irregularities in the bed and interactions between grains moving at different speeds give rise to local increases in bed elevation. These 'proto-ripples' begin as short wavelength low amplitude forms and grow to their steady state dimensions by the growth of larger forms at the expense of smaller. Each developing ripple creates a 'shadow zone' in its lee (Figure

Plate 5 Wind ripples:
(a) Ripples in medium-fine sand in the Gran Desierto Sand Sea, Sonora, Mexico.
(b) Granule ripples on the Skeleton Coast, Namibia.

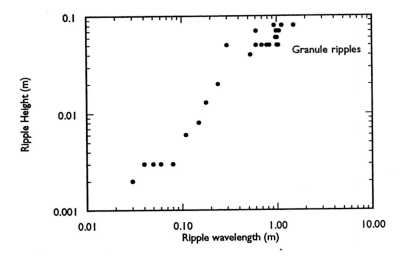

Figure 2.19 Wind ripple morphometry. Note that there is a continuum of ripple wavelength from 'normal' wind ripples to granule ripples. Data from Sharp (1963), Walker (1981), and my field observations.

2.20c), with a width proportional to ripple wavelength and impact angle. The size of the shadow zone determines the position of the next ripple downwind. Sharp argued that the controls of ripple wavelength are impact angle and ripple amplitude, both of which are dependent on grain size and wind speed, but could see no obvious reason why ripple wavelength should be dependent on the mean saltation path length. His observations on ripple development to an equilibrium size and spacing have been confirmed experimentally by the wind tunnel experiments of Seppälä and Linde (1978) and Walker (1981) and by computer simulations of sediment surfaces (Werner 1988).

Anderson (1987) has provided a rigorous model for ripple development based on experimental data and numerical simulations of sand beds. Saltating sand consists of two populations: (1) long trajectory, high impact energy 'successive saltations' and (2) short trajectory, low impact energy 'reptations'.

Figure 2.20 Models for wind ripple formation. A: Variation in impact intensity over a perturbation in the bed (after Bagnold 1941). Note higher impact intensity in zone *A–B* compared to *B–C*. B: Coincidence between ripple wavelength and mean saltation paths. C: Alternation of impact and shadow zones on a developing wind ripple (after Sharp 1963). D: Growth and movement of developing bed perturbations that evolve to wind ripples (after Anderson 1987).

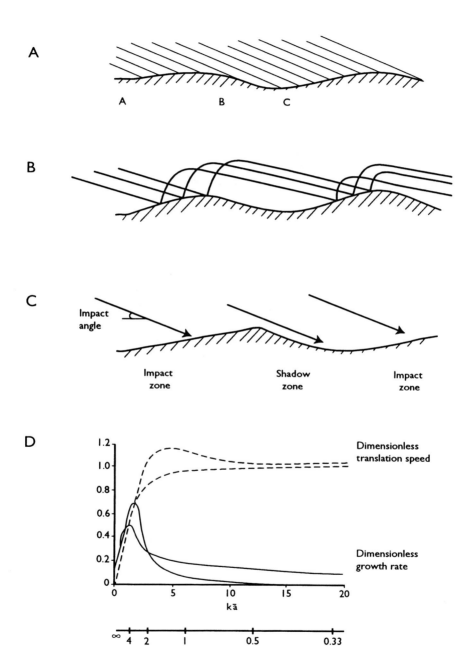

There is a wide distribution of saltation trajectories with typical path lengths that are much longer than ripple wavelengths, and a low range (1–2°) of impact angles. This suggests that the high impact energy grains do not contribute directly to ripple formation, as Bagnold hypothesized, but drive the reptation process. Using a simplified model of aeolian saltation, Anderson was able to show that a flat bed is unstable to infinitesimal variations in bed elevation, giving rise to spatial variations in the mass flux of reptating grains. Convergence and divergence of mass flux rates results in the growth of ripples, with the fastest growing ripples having a wavelength approximately 4 to 6 times the mean reptation distance (Figure 2.20d). As reptation lengths increase with wind shear stress, ripple wavelength should increase as well. This is in good agreement with wind tunnel data (Seppälä and Linde 1978; Walker 1981).

Ripple wavelengths therefore appear to scale with grain transport distance, but it is the reptation length rather than saltation path length that is the relevant parameter. The ripple pattern appears to develop from the complex merging of smaller and larger ripples with different rates of movement and by statistical fluctuations in the mass flux of reptating grains (Werner 1988; Anderson 1990). A quasi stable wavelength emerges that is the effect of the sharply decreasing rate of ripple mergers with increasing ripple size.

New models for ripple development indicate that their formation and characteristic dimensions are not directly related to the movement of grains in successive saltation, as Bagnold (1941) hypothesized. Rather, wind ripples are initiated by statistical fluctuations in the flux of reptating grains driven by impacts from a uniform distribution of saltating grains. Once the initial nuclei of the ripple pattern are set up, then the effects of local bed slope on impact intensity and the local reptation mass flux become important, as suggested originally by Sharp (1963) and confirmed by the wind tunnel experiments of Willetts and Rice (1986), and the pattern becomes self-organizing, by merging of ripples in the manner suggested by Werner (1988) and Anderson (1990).

CONCLUSIONS

Most dune sand is transported close to the surface by saltation and reptation of grains. The shearing action of the wind is resisted by grain size and shape, bed slope, moisture content, and surface roughness elements (especially the presence of vegetation). Resistances are best quantified via their effects on the threshold wind shear velocity for entrainment of sand, but also play a role when transport occurs. Modelling and wind tunnel studies show that the threshold process is complex and involves a series of interactions between the near bed airflow and the sand grains before an equilibrium is reached. In natural conditions, the natural variability of the wind as a result of turbulence and gusts suggests that such conditions will only be acheived momentarily.

The most reliable estimates of sand fluxes are derived from equations that characterize sand transport rates as a function of the ratio between wind shear velocity and the threshold wind shear velocity. The relations between flux, particle size, aerodynamic roughness, slope, wind shear velocity and actual sand fluxes on dunes are complex and may depart significantly from estimates derived from transport equations, although there are few studies of actual sand transport on desert dunes; they will be discussed in Chapter 5.

The formation of wind ripples is intimately associated with the saltation of sand. New models suggest that fluctuations in the flux of reptating grains rather than spatial patterns of saltating grains determine the initial conditions for ripple formation, which evolve rapidly by influencing the saltation impact intensity and thus the flux of reptating grains.

More than in any other area of aeolian geomorphology, studies of sand transport have been dominated by the seminal work of R. A. Bagnold. Recent developments in both computational power as well as experimental techniques have however provided new insights into the mechanisms of sand transport by the wind. The effects of boundary layer characteristics and turbulence on sand transport are increasingly being recognized, and seem likely to lead to a new appreciation of the character of sediment transport on dunes, as will be discussed further in Chapter 5.

3

DUNE MORPHOLOGY AND MORPHOMETRY

INTRODUCTION

Dunes are created by interactions between granular material (sand) and shearing flow (the atmospheric boundary layer). The resulting landforms are bedforms that are dynamically similar to those developed in sub-aqueous shearing flows (e.g. rivers, tidal currents). Their morphology reflects: (1) the characteristics of the sediment (primarily its grain size) and (2) the surface wind (both the local surface shear stress, which determines the sand transport rate as well as the directional variability of the annual wind regime). In some areas vegetation may be a significant factor, and interactions with topographic obstacles may also result in dune formation. As the dune grows upwards into the atmospheric boundary layer, the primary airflow is modified by interactions between the form and the flow which result in modification of the boundary layer flow (e.g. flow acceleration, expansion and separation, streamline convergence and divergence) and the creation of secondary flow circulations (especially in the lee of the dune). These interactions play a major role in determining dune morphology, as will be discussed in Chapter 5.

Desert dunes occur in a variety of morphologic types, each of which displays a range of sizes (height, width, and spacing). Aerial photographs and satellite images of sand seas show that most dune patterns are quite regular, and that very similar dune morphological types occur in widely separated sand seas. For example, partly vegetated linear dunes in Australia and the southwestern Kalahari are almost identical in morphology, as are compound crescentic dunes in Namibia and North America. This suggests that: (1) the local response of sand surfaces to airflow is governed by generally applicable physical principles, and (2) that there are general controls of dune size and spacing.

DUNE CLASSIFICATIONS

Many different classifications of dune types have been proposed (see Mainguet 1983, 1984a for a list of references to different schemes). They fall into two groups: (1) those that imply some relationship of dune type to formative winds or sediment supply (morphodynamic classifications), and (2) those based on the external morphology of the dunes (morphological classifications).

There are many morphodynamic classifications in which dunes are classified by their form and relation to formative winds, especially their alignment relative to the dominant or resultant (vector sum) sand transport direction (e.g. Aufrère 1928; Clos-Arceduc 1971; Wilson 1972; Hunter *et al.* 1983). Thus dunes may be classified as transverse, longitudinal or oblique (Figure 3.1), yet studies of dune dynamics (e.g. Tsoar 1983a; Lancaster 1989a) show that different parts of the same dune may be simultaneously transverse, oblique or longitudinal to the primary wind direction. Other workers (e.g. Mainguet 1983, 1984a) have attempted to order dunes by including aspects of their mobility and relation to sediment budgets and thus to distinguish between erosional types (parabolic dunes, sand ridges) and purely depositional forms (barchanoid dunes, transverse chains, linear dunes and star dunes). One problem however with all morphodynamic systems of dune classifications is that they assume a knowledge of how dunes form. As will be seen below (Chapters 5 and 6), that knowledge is at best incomplete.

The morphological classification of McKee and his co-workers (McKee 1979a) groups dunes on the basis of their shape and number of slip faces into five major types: crescentic, linear, reversing, star, and parabolic (Figure 3.2).

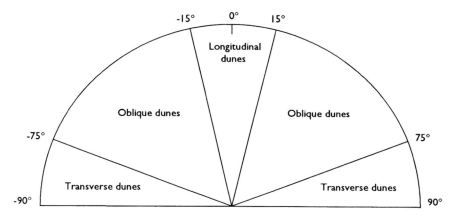

Figure 3.1 Morphodynamic classification of dunes based on the relations between dune trend and wind direction (after Hunter *et al.* 1983).

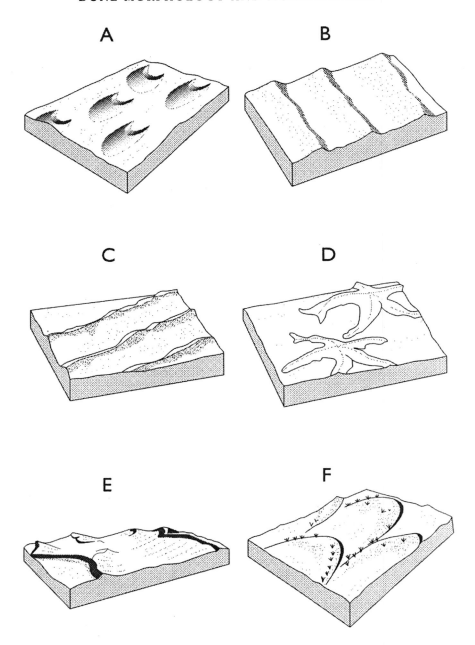

Figure 3.2 Major dune types (after McKee 1979a). A; Barchan; B: Crescentic ridges; C: Linear; D: Star; E: Reversing; F: Parabolic.

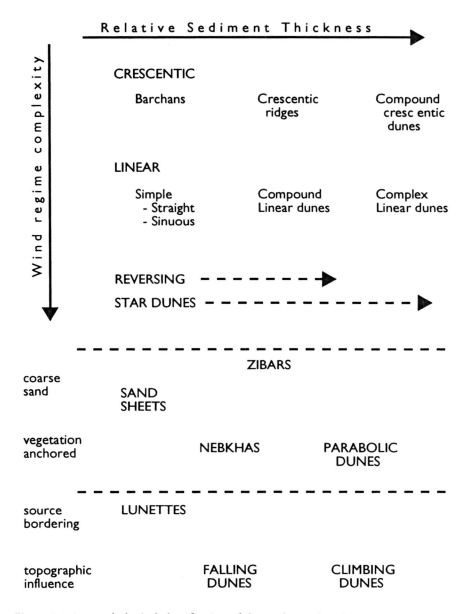

Figure 3.3 A morphological classification of desert dunes (based in part on McKee 1979a and Pye and Tsoar 1990).

In turn, three varieties of each dune type can occur: simple, compound and complex. Simple dunes are the basic form of each dune type. Compound dunes are characterized by superimposition or juxtaposition of dunes of the same morphological type (e.g. superimposition of smaller crescentic dunes on the stoss side of large crescentic dunes). Complex dunes occur where dunes of two types are superimposed or merged (e.g. crescentic dunes on the flanks of larger linear or star dunes, or linear dunes with star dune peaks). Compound or complex dunes are a common feature of many modern sand seas and comprise 46.6 per cent of the dunes in sand seas examined by Fryberger and Goudie (1981). They are equivalent to the draa of Wilson (1972), and this term has been applied to all large dunes with superimposed bedforms (Kocurek 1981).

In addition to the major dune types identified above, many sand seas and dune fields contain areas of gently undulating to flat sand surfaces (sand sheets), and areas of low rolling dunes without slip faces (zibar). Interactions between vegetation and sand accumulations lead to the formation of shadow dunes (also called shrub-coppice dunes or nebkha) (Nickling and Wolfe 1994). Where sand accumulates adjacent to topographic obstacles, echo dunes or climbing dunes occur on their windward side, and lee dunes or falling dunes on the lee side.

Most dune types can be accommodated in an expanded morphological classification of dunes, following Pye and Tsoar (1990). My version of this classification is presented in Figure 3.3, and will be the basis for this book, although it is not intended to be a comprehensive model.

AEOLIAN BEDFORM HIERARCHIES

In all sand seas and dune fields, there is a hierarchical system of aeolian bedforms superimposed on one another (Matschinski 1952; Wilson 1972; Lancaster 1988a). Similar bedform hierarchies occur in sub-aqueous environments (Allen 1968a; Jackson II 1975). Wind ripples cover at least 80 per cent of sand surfaces in all dune areas, whereas in many sand seas large dunes (compound or complex dunes) are characterized by the development of smaller dunes superimposed on their stoss or lee slopes. Three orders of aeolian bedforms can therefore be identified (Figure 3.4), although only the first two occur in all sand seas: (1) wind ripples (spacing 0.1–1 m), (2) individual simple dunes or superimposed dunes on compound and complex dunes (spacing 50–500 m), and (3) compound and complex dunes or draa (spacing > 500 m).

A range of sizes of dunes occurs in all desert sand seas, yet satellite images and air photographs of desert sand seas show that most dune patterns are very regular. Regular, ordered spacing of dunes of a limited range of sizes in a given area gives rise to the close relationships between dune height, width, and spacing that have been documented by many workers (e.g. Wilson 1972;

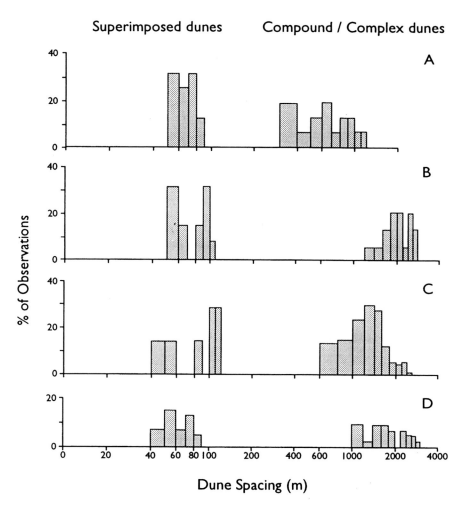

Figure 3.4 The hierarchy of aeolian bedforms of superimposed dunes and compound and complex dunes. A: Compound crescentic dunes (Namib); B: Compound and complex linear dunes (Namib); C: Compound crescentic dunes (Gran Desierto; D: Compound crescentic (Gran Desierto) (after Lancaster 1988a).

Breed and Grow 1979; Lancaster 1983a; Wasson and Hyde 1983a; Lancaster *et al.* 1987; Thomas 1988) for widely separated sand seas (Figure 3.5). This organization of dunes into regular patterns is a classic example of self-organization in a geomorphic system (Hallet 1990).

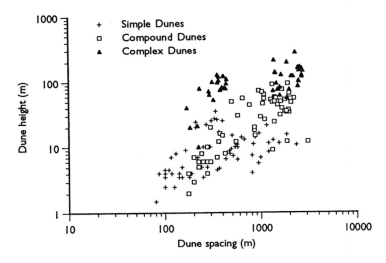

Figure 3.5 Relations between dune height and spacing. Note overlap between simple, compound, and complex dunes in this global sample (from Lancaster 1988a).

MORPHOLOGY AND MORPHOMETRY OF MAJOR DUNE TYPES

Crescentic dunes

The simplest dune types and patterns are those that form in wind regimes characterized by a narrow range of wind directions. In the absence of vegetation, crescentic dunes will be the dominant form. Hunter *et al.* (1983) and Tsoar (1986) indicate that this dune type is stable where the directional variability is 15° or less about a mean value. Isolated crescentic dunes or barchans occur in areas of limited sand availability. As sand supply increases, barchans coalesce laterally to form crescentic or barchanoid ridges that consist of a series of connected crescents in plan view (McKee and Douglass 1971; Kocurek *et al.* 1992). Larger forms with superimposed dunes are termed compound crescentic dunes (e.g. Breed and Grow 1979; Havholm and Kocurek 1988; Lancaster 1989b). Complex crescentic dunes occur in some sand seas, especially where star or reversing dunes develop from superimposed crescentic forms (e.g. Lancaster *et al.* 1987; Havholm and Kocurek 1988).

Figure 3.6 Terminology for dune morphology. A: Barchan; B: Crescentic ridge; C: Linear dune; D: Star dune.

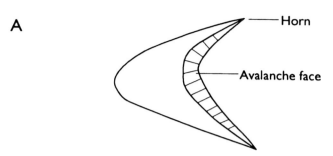

A

Horn

Avalanche face

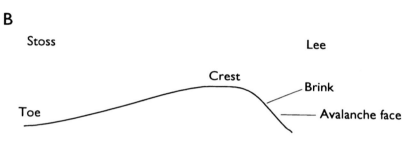

B

Stoss

Lee

Crest

Brink

Toe

Avalanche face

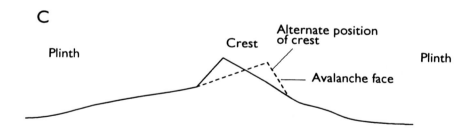

C

Alternate position
of crest

Plinth

Crest

Plinth

Avalanche face

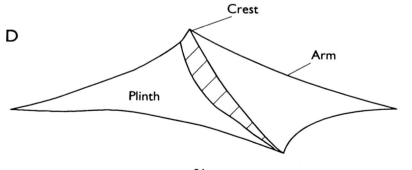

D

Crest

Arm

Plinth

Barchans

Barchans (Plate 6a, Figure 3.6a) occur in two main areas: (1) on the margins of sand seas and (2) in sand transport corridors linking sand source zones with depositional areas. Despite the attention that has been given to this dune form, they comprise a very small proportion of global sand deposits (Fryberger and Goudie 1981). Small barchans occur on the northern margins of the Namib Sand Sea (Slattery 1990) and on the downwind margins of the Algodones and Skeleton Coast dune fields (Smith 1980; Sweet *et al.* 1988; Lancaster 1982a). Barchans are also found on the upwind margins of the Namib Sand Sea (Lancaster 1989b), the Skeleton Coast dune field, and at White Sands (McKee 1966). The most widespread development of barchans occurs in sand transport corridors extending downwind from source zones, such as in the Kharga Oasis in Egypt (e.g. Bagnold 1941; Embabi 1982), the Jafurah Sand Sea (Saudi Arabia) (Fryberger *et al.* 1984), the Pampa la Joya in Peru (Finkel 1959; Hastenrath 1967) and the southern Namib (Endrody-Younga 1982; Corbett 1993). Most areas of barchans seem to be characterized by a gravel- or granule-covered substrate and an irregular spacing of dunes.

Wind regimes in areas of barchans (Figure 3.7a) are characterized by the dominance of one directional sector, although seasonal reversal of wind directions involving short-lived high intensity storms does occur (e.g. in the Namib). Barchans appear to be a very stable form in areas of limited sand supply and unidirectional winds, as demonstrated by their ability to migrate long distances with only minor changes of form (e.g. Long and Sharp 1964; Hastenrath 1987; Haynes 1989). For example, some fifty years later, Haynes (1989) was able to recognize the same barchan that R. A. Bagnold and his colleagues camped next to in 1930. In this period the dune had migrated 375 m downwind. In the Salton Sea area of California, the same barchans can be identified and followed on aerial photographs spanning a thirty-year interval (Long and Sharp 1964; Haff and Presti 1984).

Barchans are characterized by an ellipsoidal shape in plan view, with a concave slip face and 'horns' extending downwind (Plate 6a, Figure 3.6a). Most barchans range in height between 3 and 10 m. Very large barchans (mega barchans) with superimposed crescentic dunes have been described from a few localities, including the 55 m high Pur Pur dune in Peru (Simons 1956), the southern end of the Algodones Dunes, the Snake River Plain in Idaho, and the Skeleton Coast dune field (Lancaster 1982a). The stoss slope of barchans is convex in profile, with slope angles of 2–10° (Hastenrath 1967; Tsoar 1985), and the crest may lie some distance upwind of the brink of the avalanche face. Dune height is typically about 1/10 of the dune width (Finkel 1959; Hastenrath 1967; Embabi 1982). Lower barchans appear to have a generally flatter profile. Strongly elongated horns and asymmetric development of barchan plan shapes occurs in some areas (e.g. Hastenrath 1967; Lancaster 1982a), and has been attributed to asymmetry in the wind regime

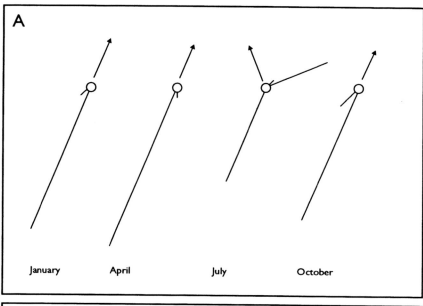

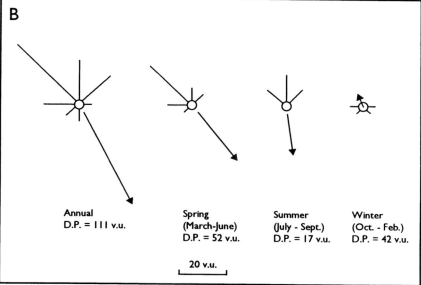

Figure 3.7 Wind regimes of crescentic dunes. A: Simple and compound crescentic dunes, Namib Sand Sea. B: Compound crescentic dunes, Algodones dunes, California (data from Havholm and Kocurek 1988, Figure 3).

or sand supply. In some areas, barchans of this type are transitional to linear dunes (Lancaster 1980; Tsoar 1984).

Crescentic ridges

Crescentic ridges of simple and compound varieties occupy about 40 per cent of the area of sand seas world wide (Fryberger and Goudie 1981), and occur in all desert regions. Crescentic ridges are also called barchanoid ridges or transverse dunes by many investigators. They are the dominant dune form in the Thar Desert, the Takla Makan, and Tenegger sand seas of Asia; the Jafurah, Nafud, and eastern and northern Rub' al Khali of Saudi Arabia; and in the northern Saharan sand seas (Fryberger and Goudie 1981). In southern Africa, crescentic dunes occupy some 13 per cent of the area of the Namib Sand Sea, and are the main dune type in the smaller northern Namib dune fields. They are also very widespread to dominant in all North American dune fields.

(a)

Simple crescentic ridges

The patterns of most simple crescentic ridges are quite regular, as indicated by the close correlations that exist between dune height, width and spacing (Figure 3.8). Typical areas of simple crescentic ridges occur at White Sands, New Mexico (McKee 1966), Guerro Negro, Baja California (Inman *et al.* 1966), on the east margin of the Algodones dune field (Sweet *et al.* 1988), and in the Skeleton Coast dune field in the northern Namib (Lancaster 1982a). Their morphometry is summarized in Table 3.1. Simple crescentic ridges occur in areas of unidirectional wind regimes similar to those found in areas of barchan dunes (Figure 3.7).

In the Skeleton Coast dune field, most simple crescentic and barchanoid ridges (Plate 6b) are between 3 and 10 m high, with a spacing of 100–400 m (Lancaster 1982a). In profile, windward slopes steepen from 2–3° at the base to 10–12° in the upper mid slope and decline again to 2–3° at the crest, which is generally sharp, but occasionally rounded, with the brink of the slip face, at an angle of 32–34° lying beyond the crestline (Figure 3.6b). In plan view, some crescentic ridges (Plate 6b) are straight or very gently curved, and are called 'transverse ridges' by McKee (1979a). Other crescentic ridges are

(b)

Plate 6 Simple crescentic dunes:
(a) Barchans on the upwind margin of the Skeleton Coast dune field, Namibia.
(b) Simple crescentic ridges, Skeleton Coast dune field, Namibia.

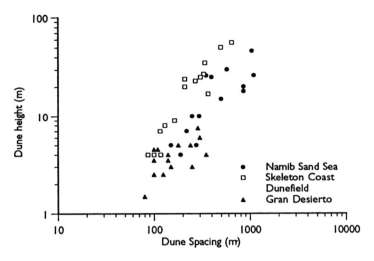

Figure 3.8 Relations between dune height and spacing, crescentic dunes.

Table 3.1 Morphometry of simple crescentic dunes. Mean values, range in parentheses.

Location	Spacing (m)	Width (m)	Height (m)
White Sands	112	72	
	(60–198)	(30–185)	
United Arab Emirates	590	440	
	(70–2000)	(100–1100)	
Al Jiwa	630	590	
	(200–1000)	(200–900)	
Thar Desert	580	470	
	(200–1000)	(500–1500)	
Takla Makan	590	630	
	(500–1200)	(200–1000)	
Gran Desierto	150–400		
Namib Sand Sea	272		8.25
	(100–400)		(3–10)
Skeleton Coast	259	21	
	(90–645)	(4–56)	

Source: Data from Breed and Grow (1979) supplemented by my field measurements.

sinuous in plan and consist of a series of connected crescents that form the higher parts of the dune ridge separated by lower saddle areas. These were called barchanoid and linguoid elements by Wilson (1972), and this variant of a crescentic ridge is also known as a barchanoid ridge. In some areas of

56

complete sand cover where the toe of one dune lee face abuts the base of the stoss slope of the next dune downwind, linear ridges cross interdune areas from one linguoid element to the next giving rise to enclosed interdune hollows and a rectilinear or aklé pattern (Cooper 1958; Cooke and Warren 1973). In the Skeleton Coast dune field, straight crested dunes occur on the western side where sand supply is greater; and barchanoid ridges are more common in the east where sand supply is less and the sub-dune surface more irregular.

Compound crescentic dunes

Compound crescentic dunes (Plates 2 and 7) are characterized by a major dune ridge that rises to a height of 20–80 m above the interdune area, with multiple crescentic dunes superimposed on the upper stoss and crestal areas. The spacing of the major dune ridges ranges from around 700 to as much as 2000 m (Table 3.2). Compound crescentic dunes are particularly common in the Takla Makan and northern Saharan sand seas (50 per cent of all dunes) and are important in the Nafud and northern Rub' al Khali of Saudi Arabia (Fryberger and Goudie 1981). Dunes of this type comprise about 10 per cent of the Gran Desierto and Namib sand seas (Lancaster *et al.* 1987; Lancaster

Plate 7 Compound crescentic dunes in the Algodones dune field, California.

57

Table 3.2 Morphometry of compound crescentic dunes. Mean values, range in parentheses.

Location	Spacing (m)	Width (m)	Height (m)
Gran Desierto	1380 (500–2300)	660 (300–1500)	20–100
Algodones	1070 (400–2500)	880 (500–2500)	50–80
Nafud	1840 (800–3300)	800 (500–2000)	
Rub' al Khali	1430 (850–2200)	670 (300–1100)	
Thar Desert	1440 (700–2500)	1300 (750–2000)	
Takla Makan	3000 (2000–5500)	2200 (1100–3400)	
Aoukar	1710 (1000–2500)	1590 (1200–2100)	
NW Sahara	650 (300–1500)	1240 (500–2000)	
Namib	694 (800–1200)	680 (300–1200)	18.6 (10–40)

Source: Data from Breed and Grow (1979) supplemented by my field measurements.

1989b) and more than 50 per cent of the Algodones dune field (Sweet *et al.* 1988). Typical areas of compound crescentic dunes are the coastal areas of the Namib Sand Sea, the Algodones dune field of southern California, and the eastern parts of the Gran Desierto Sand Sea.

In the Namib, this dune type occurs in a strip up to 20 km along the Atlantic coast (Lancaster 1989b). The dunes there consist of a main dune ridge 20–50 m high with a spacing of 800–1200 m that is aligned transverse to S to SSW winds (Plate 2). There are no true interdune areas, and the base of the slip face of one dune abuts the lower stoss slopes of the next dune downwind. Crescentic ridges, 2–5 m high with a spacing of 50–100 m, are superimposed on the upper stoss slopes and crestal areas of the main dunes. Compound crescentic ridges in the Gran Desierto Sand Sea are 300–1000 m wide and 20–50 m high (Plate 6a). Dune spacing ranges from 1400 m to as much as 2000 m. Superimposed on the stoss (southern) side of the main ridges, which are aligned approximately transverse to S–SE winds, are multiple 2–5 m high crescentic ridges with a spacing of 50–100 m. In the centre of the main ridge there are large, crescentic avalanche faces oriented towards the north. Compound crescentic dunes in the Algodones dune field (Plate 7) are up to 80 m high with straight to gently curved lee faces oriented to the south. In the southern third of the dune field deflated gravel-covered flats up to 500 m wide separate 60 m high dunes (Sharp 1979; Havholm and

Kocurek 1988; Sweet *et al.* 1988). Superimposed crescentic dunes up to 20 m high occur on the stoss, crest, and some lee slopes of the main dunes. Superimposed dune height tends to increase toward the main dune crest.

Wind regimes in areas of compound crescentic dunes are generally similar to those for other varieties of crescentic dunes (Figure 3.7). However, the Algodones dunes are maintained by a wind regime that has three main directional sectors: summer winds (late June to August) are southerly, winter winds (late October to February) are from the north, and spring winds (March to early May) are from the west and northwest. The vector sum of sand transport is toward the southeast, or approximately perpendicular to the main dune crestline (Havholm and Kocurek 1988).

Complex crescentic dunes

Complex crescentic dunes, some with star dunes superimposed on their crests, have been described from several sand seas, including the Gran Desierto, Algodones dune field, Grand Erg Occidental, Awbari and Murzuq sand seas of Libya, and the Thar Desert (Breed *et al* 1979; Kar 1990). In some areas, these dunes are transitional to star dunes, as crescentic dunes migrate into areas of more complex wind regimes (Lancaster 1989a).

Linear dunes

Linear dunes (Figure 3.6c) are characterized by their length (often more than 20 km), straightness, parallelism and regular spacing, and high ratio of dune to interdune areas. Lancaster (1982b) estimated that 50 per cent of all dunes are of linear form, with the percentage varying between 85–90 per cent for areas of the Kalahari and Simpson-Strzelecki Deserts to 1–2 per cent for the Ala Shan and Gran Desierto Sand Seas. Linear dunes are the dominant form in sand seas in the southern hemisphere and in the southern and western Sahara.

There are several varieties of linear dunes. Simple linear dunes are of two types: the long, narrow, straight, partly vegetated linear dunes of the Simpson and Kalahari Deserts (the vegetated linear dunes of Tsoar (1989) (Plate 8a)), and the more sinuous sharp-crested dunes often called seifs (Plate 8b) found in Sinai (Tsoar 1983a), parts of the eastern Sahara (Bagnold 1941), and along the west edge of the Algodones dune field (Sweet *et al.* 1988). Compound linear dunes (Plate 9) consist of two to four sinuous ridges on a broad plinth and are typified by those in the southern Namib Sand Sea and the Fachi Bilma Erg (Mainguet and Callot 1978; Lancaster 1983b). Large (50–150 m high, 1–2km spacing) complex linear dunes (Plate 10) with a single sinuous main crestline, distinct star-form peaks, and crescentic dunes on their flanks, occur in the Namib and Rub' al Khali Sand Seas (Holm 1960; Lancaster

(a)

(b)

Plate 8 Simple linear dunes:
(a) Vegetated linear dunes in the northern Gran Desierto, Mexico.
(b) Sinuous crested dunes in the northern Namib Sand Sea.

Table 3.3 Morphometry of linear dunes. Mean values, range in parentheses.

Location	Spacing (m)	Width (m)	Height (m)
Simple			
Simpson[1]	648	290	11.4
	(431–1148)		(6.5–21.0)
Great Sandy Desert[1]	1134		9.09
	(370–2346)		(4.1–15.4)
SW Kalahari[2]	435	220	9.0
	(431–1148)		(2–20)
Compound			
Namib[3]	1724	650	34.5
	(990–2082)		(24–48)
SW Sahara[4]	1930	940	
Complex			
Namib[3]	2163	880	99.5
	(1500–2760)		(44–167)
Rub' al Khali[4]	3170	1480	100–200
S Sahara[4]	3280	1280	

Sources: Data from: (1) Wasson and Hyde 1983a; (2) Lancaster 1988b; (3) Lancaster 1989b; (4) Breed and Grow 1979.

1983b). Wide (1–2 km) complex linear dunes with crescentic dunes super-imposed on their crests occur in the eastern Namib, parts of the Wahiba Sands (Warren 1988) and the Akchar Sand Sea of Mauritania (Kocurek et al. 1991). Table 3.3 summarizes the morphometry of linear dunes.

Simple linear dunes

Simple linear dunes of the straight, partly vegetated variety are typified by those in the Simpson-Strzelecki and southwestern Kalahari sand seas. In both areas, the dominant dune type consists of straight to slightly sinuous, parallel to sub-parallel ridges (Plate 8a, Figure 3.9a), which are 2–35 m high, with a spacing of 200–450 m and a width of 150–250 m (Lewis 1936; Folk 1970; Lancaster 1988b; Wasson et al. 1988). Most dunes are 20–25 km long, although individual examples have been traced for over 200 km (Madigan 1946). Other areas of simple, vegetated linear dunes are found on the Moenkopi Plateau of northeastern Arizona (Hack 1941; Breed and Grow 1979), in parts of Sinai and Mojave Deserts (Tsoar and Møller 1986), and on the northern margins of the Gran Desierto (Lancaster et al. 1987). Several varieties of simple linear dune patterns have been identified in both the Simpson Desert and the Kalahari (Mabbutt and Wooding 1983; Thomas 1986).

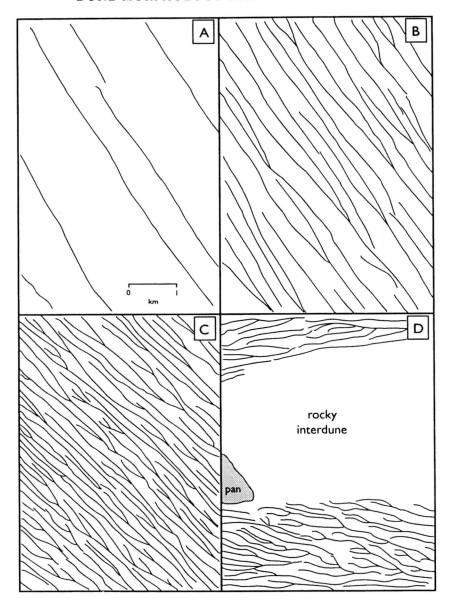

Figure 3.9 Patterns of simple and compound linear dunes in the southwestern Kalahari. A: Broad widely spaced ridges in the N and NE parts of the dune field; B: Straight parallel ridges with 'Y' junctions in the central part of the dune field; C: Narrow closely spaced ridges with common 'Y' junctions in the SE part of the dune field; D: Clustered dendritic ridges (compound linear dunes) west of the lower Molopo river. (After Lancaster 1988b).

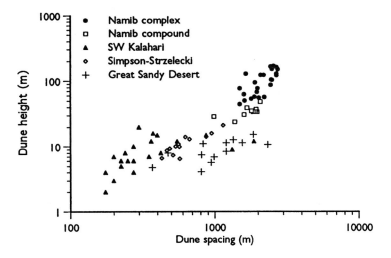

Figure 3.10 Height/spacing relations for linear dunes. Data for Australian areas from Wasson and Hyde 1983a.

In the Kalahari and Simpson Deserts, each dune ridge consists of a broad plinth, with a slope of 2–4°, rising to a crest line which consists of a series of peaks and saddles with a local relief of 1–3 m. Upper flanks of the dunes have slope angles of 5–18°. In profile the dunes are asymmetric, with steeper slopes on their southwest or south sides (13.4°) compared to their northeast or north sides (8.4°) in the Kalahari, and a gentle western slope (12°) and a steeper eastern slope (20°) in Australia. Avalanche faces are poorly developed and seldom exceed 2 m in height. A characteristic feature of areas of this dune type are 'Y' or tuning-fork junctions open to the upwind direction (Figure 3.9). 'Y' junctions are more common in areas of closely spaced dunes (Mabbutt and Wooding 1983; Thomas 1986), and are one mechanism by which the dune pattern adjusts its spacing to changed conditions of sediment availability or substrate type.

Interdune areas between simple linear dunes are often relatively well vegetated and may be sand-covered or reflect the substrate over which the dunes pass: locally alluvial deposits or gibber (desert pavement). Mabbutt and Wooding (1983) and Wasson *et al.* (1988) have shown that dune spacing tends to increase as dunes pass into areas of gravel substrate on the margins of the dune fields. Closely spaced dunes are associated with areas of fine-grained interdune sediments. The height and spacing of dunes between blocks of linear dunes in both deserts are closely correlated (Figure 3.10) although there is considerable scatter within each area (Wasson and Hyde 1983a; Lancaster 1988b; Thomas 1988).

Compound linear dunes

Compound linear dunes are best documented from the southern parts of the Namib Sand Sea (Plate 9, Figure 3.11d) where they consist of three and locally up to five slightly sinuous 5–10 m high sharp crested ridges running parallel or sub-parallel to each other on SE–NW alignments (Lancaster 1983b). These dunes are very similar to the 'bouquets de silks' and 'silks sur ondulations' described by Mainguet and Callot (1978) from the Erg Fachi Bilma. Forms comparable with compound linear dunes are the broad crested linear dunes of the Great Sandy Desert (Wasson *et al.* 1988) and the coalesced or clustered dendritic linear dunes of the southwestern Kalahari (Goudie 1970; Thomas 1986; Lancaster 1988b) (see Figure 3.9d).

In the Namib, the crestal ridges lie on a 500–800 m wide gently convex linear ridge, composed of coarser sand and with the same general alignment as the crestal ridges and similar to the 'whale backs' of Bagnold (1941) and the 'ondulations longitudinales' of Mainguet and Callot (1978). Total height of these dunes is 25–50 m and they have a spacing of 1200–2000 m. Locally the crestal ridges may join in 'Y' junctions open to the south or southeast. In other places, dendritic or 'feather barb' patterns may occur. The ridges are asymmetric, with the aspect of the slip faces changing seasonally from SW or W in April to September and NE or E during the remainder of the year. Windward slopes of the crestal ridges are between 15 and 25°. The lightly

Plate 9 Compound linear dunes in the southern Namib Sand Sea.

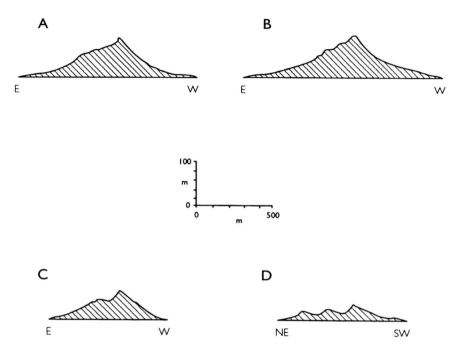

Figure 3.11 Cross-sections of different varieties of large linear dunes in the Namib (after Lancaster 1989b). A–C: complex varieties, D: compound variety.

vegetated lower slopes, or plinths, on each side of the dune slope at 2–6° steepening towards the centre of the dune. Interdune areas between compound linear dunes of this type are generally sand covered.

Complex linear dunes

Complex linear dunes occur in the Namib Sand Sea (Lancaster 1983b), the Rub' al Khali (Breed *et al.* 1979), the northern Sahara in Algeria, and the Akchar Erg of Mauritania (Kocurek *et al.* 1991). Those in the Namib Sand Sea consist of a single main dune ridge on a S–N to SSE–NNW alignment which rises to 50–170 m above adjacent interdune areas (Plate 10, Figure 3.11). Individual dunes are spaced 1600–2800 m apart, with a mean of 2108 m. The overall dune pattern is regular, with a close correlation between dune height and spacing (Figure 3.10). The crestline of the main ridge is sharp and sinuous and connects a series of regularly spaced peaks. The major slip face, at an angle of 32–33°, faces east or northeast and may be 10 m high at the time of its maximum development in March. In winter, east to northeast winds

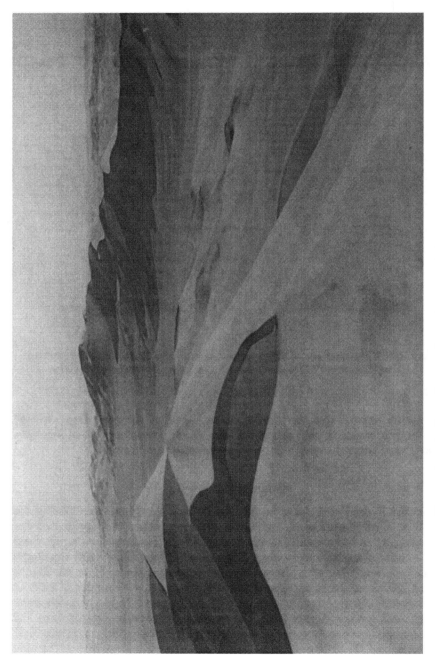

Plate 10 Complex linear dunes in the northern part of the Namib Sand Sea.

erode its upper section and reverse the orientation of the slip face to face west or southwest. However, this slip face is rarely more than 5 m high (Livingstone 1989). Below the slip faces on the eastern side of the dune is a wide, sparsely vegetated, gently sloping plinth. Secondary or superimposed dunes of barchanoid or crescentic form, 2–10 m high and 50–200 m apart, are developed on the upper parts of the plinth (Livingstone 1987a). Strong development of east flank barchanoid dunes is associated with a sinuous main crest. Where it is straight, or slightly sinuous, such dunes are absent or poorly developed. Western slopes and plinths of the linear dunes are smooth or gently undulating in a direction normal to the main dune trend. Slope angles steepen from 2–5° on the plinth to 15–20° near the crest. Most interdune areas between complex linear dunes in the Namib are sand covered with undulations on a trend normal or slightly oblique to that of the main dune ridges. In places, these undulations continue onto the plinths and upper western slopes of the adjacent linear dunes.

Complex linear dunes in the Akchar Erg of Mauritania are commonly between 15 and 30 m high, but may reach 75 m above the interdune areas. They have a crest-to-crest spacing of 200 m. Superimposed on the linear ridges are 2–3 m high crescentic dunes, oriented approximately perpendicular to the trend of the linear features (Kocurek et al. 1991). Complex linear dunes in the Wahiba Sands described by Warren (1988) are up to 100 m high, 1200 m wide and have a spacing of 2000 m. Reticulate or 'network' patterns of crescentic dunes occur on their broad crests and narrow simple linear dunes in interdune corridors and dune flanks.

Wind regimes of linear dunes

Linear dunes are found in a range of wind regimes both of total energy and directional variability (Figure 3.12). They are commonly found in wide unimodal (winds from one broad directional sector) or bi-directional wind regimes (winds from two distinct directions) and occasionally occur in complex wind regimes (more than two modes) (Fryberger 1979).

Regional-scale systems of linear dunes have been described from the Sahara (Mainguet and Canon 1976; Fryberger 1979), Australia (Figure 3.13) (Jennings 1968; Wasson et al. 1988), and southern Africa (Figure 3.14) (Lancaster 1981a; Thomas and Shaw 1991a). This type of dune system forms large arcs corresponding approximately to the pattern of outblowing winds around anticyclonic cells, but influenced also by winds associated with temperate cyclones on their poleward sides, especially in Australia, and the Intertropical Convergence Zone (ITCZ) on their equatorial margins (Fryberger 1979). The development of linear dunes appears to be favoured by the wind regimes associated with this type of circulation with relatively persistent winds from one major direction, together with seasonally important cross-winds.

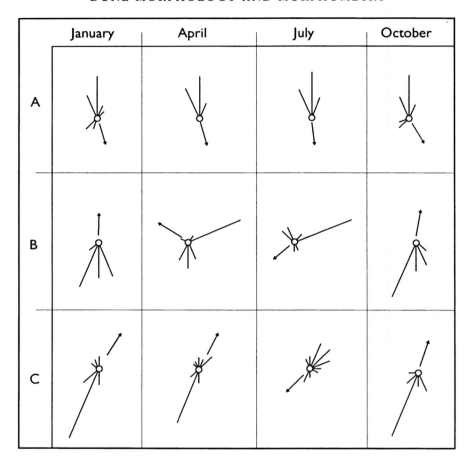

	January	April	July	October
A				
B				
C				

Figure 3.12 Wind regimes of linear dunes. A: Simple, vegetated: Upington, South Africa; B: Compound: Harlenberg, Namibia; C: Complex: Narabeb, Namibia.

Wind regimes in areas of simple vegetated linear dunes are typified by that of Oodnadatta in the Simpson Desert of Australia and Upington in the Kalahari (Figure 3.12a) (Fryberger 1979). In these areas, winds change seasonally over a range of directions on each side of the dune trend so that over the year, the resultant or vector sum transport direction is approximately parallel to the dunes. Some linear dunes occur in areas where there is a clearly bi-directional wind regime. Tsoar (1983a) identified two sectors of sand-moving winds in the Sinai: southerly winds associated with cyclonic circulation in winter and northwesterly sea breezes in summer months.

Wind regimes in areas of complex linear dunes are more variable compared

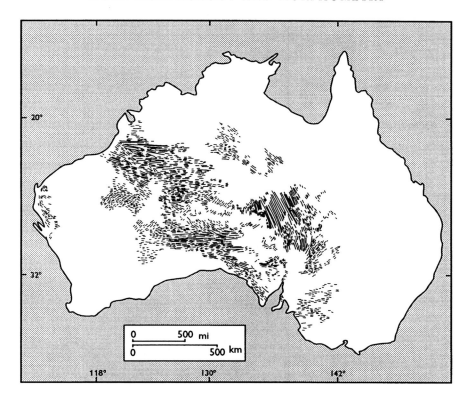

Figure 3.13 Linear dune systems in Australia (after Jennings 1968).

to simple and compound varieties (Fryberger 1979), but one or more adjacent directions are persistent, and these appear to determine the overall alignment of the dunes (Figure 3.12c). In some cases, winds from directions perpendicular to the dunes form an significant subsidiary component of the wind regime. In the Namib Sand Sea two main directional sectors are important in areas of complex linear dunes: SW–SSW and NE–E. The southwesterly winds are the product of the sea breeze and are most persistent and strongest during summer months (September to April) and generate some 40 per cent of annual potential sand movement. Strong easterly to northeast winds occur during the winter months and are responsible for 25–30 per cent of sand flow. Thus there are two major sand-moving winds from sectors 130° apart, but the southwesterly sector dominates, giving a net sand movement at an angle of 20–30° to the trend of the dunes.

69

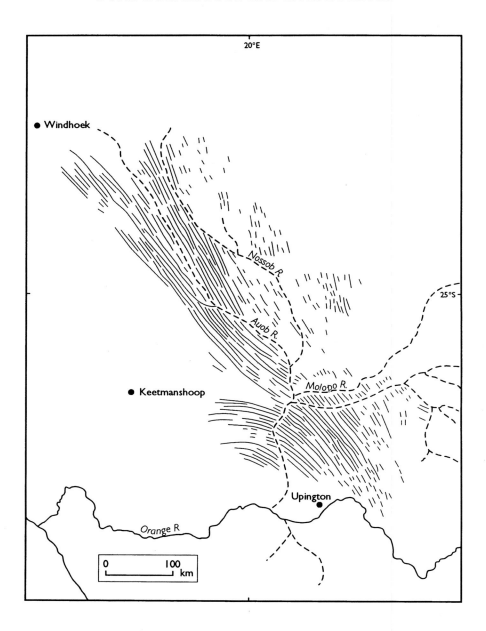

Figure 3.14 Linear dune systems in the SW Kalahari (modified from Lancaster 1988b).

Star dunes

Star dunes (Plate 11, Figure 3.6d), characterized by their large size, pyramidal morphology and radiating sinuous arms, are the largest dunes in many sand seas and may reach heights of more than 300 m. They contain a greater volume of sand than any other dune type (Wasson and Hyde 1983b) and occur in areas which represent depositional centres in many sand seas (Mainguet and Callot 1978; Lancaster 1983a).

Approximately 8.5 per cent of all dunes are of star type, and occur in many sand seas, including the Grand Erg Oriental in Algeria; the Ala Shan Desert in China; the southeastern Rub' al Khali in Arabia; the Gran Desierto in Mexico; the Erg Fachi-Bilma in Niger and the Namib Sand Sea in southern Africa (Breed *et al.* 1979; Breed and Grow 1979; Capot-Rey 1947; Holm 1960; McKee 1966, 1982; Mainguet and Callot 1978; Lancaster 1983a; Mainguet and Chemin 1984; Lancaster *et al.* 1987). In the Basin and Range province of the North American desert there are many occurrences of star-like dunes (Sharp 1966; Andrews 1981; Nielson and Kocurek 1987).

The area covered by star dunes varies considerably from one sand sea to another, but is rarely more than 10 per cent. Star dunes are absent in Australian, Kalahari and Indian sand seas, as well as those in the eastern, western and southern Sahara. They cover 9 to 12 per cent of the dune area of sand seas in the northern Sahara, central Asia, the Namib and the Gran Desierto (Fryberger and Goudie 1981; Lancaster 1983a; Lancaster *et al.* 1987). The only sand sea in which star dunes cover a large proportion of its area is the Grand Erg Oriental in Algeria, where 40 per cent of dunes are of this type (Breed *et al.* 1979; Mainguet and Chemin 1984).

In many sand seas, complex dunes (McKee 1979a) are transitional to star dunes. Star dunes may develop from complex linear dunes in the Namib Sand Sea (Lancaster 1983a) and complex crescentic dunes in the Gran Desierto, northern Saharan, Arabian and central Asian sand seas (Breed *et al.* 1979; Lancaster *et al.* 1987; Walker *et al.* 1987).

Star dunes are characterized by a pyramidal shape, with three or four arms radiating from a central peak and multiple avalanche faces. Each arm has a sharp sinuous crest, with avalanche faces which alternate in aspect as wind directions change seasonally. The arms may not all be equally developed and many star dunes have dominant or primary arms on a preferred orientation. Star dunes in the Gran Desierto Sand Sea have primary arms oriented approximately transverse to the main wind directions in the area (northerly and southeasterly) (Lancaster 1989a). Deep hollows often occur between the arms. The upper parts of many star dunes are very steep with slopes at angles of 15–30°. Avalanche face orientation changes seasonally. The basal parts of star dunes consist of a broad, gently sloping (5–10°) plinth or apron. Small crescentic or reversing dunes are often superimposed on the lower flank and upper plinth areas of star dunes.

(a)

(b)

Plate 11 Star dunes:
(a) At Sossus Vlei, Namibia.
(b) Gran Desierto.

Table 3.4 Morphometry of star dunes in different sand seas. Mean values, range in parentheses.

Location	Spacing (m)	Width (m)	Height (m)
Namib[1]	1330	1000	145
	(600–2600)	(400–1000)	(80–350)
Niger[2]	1000	610	
	(150–3000)	(200–1200)	
Grand Erg Oriental[2]	2070	950	117[3]
	(800–6700)	(400–3000)	
SE Rub' al Khali[2]	2060	840	
	(970–2860)	(500–1300)	(50–150)[4]
Gran Desierto[5]	2982	2092	
Clusters	(1500–4000)	(700–6000)	
Dunes in clusters	312	183	80
	(160–488)	(90–363)	(10–150)
Ala Shan[2]	137	740	
(Badain Jaran Shamo)	(300–3200)	(400–1000)	(200–300)[6]

Sources: Data from: (1) Lancaster (1989b); (2) Breed and Grow (1979); (3) Wilson (1972); (4) Holm (1960); (5) Lancaster et al. (1987); (6) Walker et al. (1987).

The spacing of star dunes in the global sample of occurrences studied by Breed and Grow (1979) ranges from 150 m to more than 5000 m (Table 3.4): most dunes have a spacing between 1000 and 2400 m. The available data on the heights of star dunes indicate that they are in many cases the highest dunes in a sand sea. Many of the claims for the 'world's highest dunes' relate to star dunes. In the Grand Erg Oriental, star dunes are 230 m high in its southwestern areas (Breed et al. 1979), and average 117 m in height (Wilson 1973). Wilson (1973) reports that star dunes in the Issaouane-N-Irrararene (Algeria) are 300–430 m high. In the Badain Jaran sand sea in China, Walker et al. (1987) report star dunes 200–300 m high, with some reaching as much as 500 m. In the north American deserts, star dune peaks at Kelso are 170 m high (Sharp 1966), 122 m at Dumont, and 203 m at Eureka. The mean height of star dunes in the Namib Sand Sea is 145 m. The largest star dunes occur in the vicinity of Sossus Vlei, where they attain heights of 300–350 m. Those in the southern parts of the sand sea are much lower, at 80–150 m, and more closely spaced (± 1000 m). In the Gran Desierto, many star dunes are 80–100 m high, with a maximum of 150 m. Star dune height and spacing are correlated in both the Namib and Gran Desierto Sand Seas, but the relation is much less strong than for crescentic and linear dunes (Figure 3.15).

Star dunes in the Namib Sand Sea (Plate 11a) consist of a single narrow, steep sided ridge with a straight or sinuous crestline, which is near symmetrical in profile and has a SE–NW or SSE–NNW alignment. From the crestal ridge, curving arms descend on alignments which are roughly perpendicular to the crest. Slip faces develop on both sides of the crest and

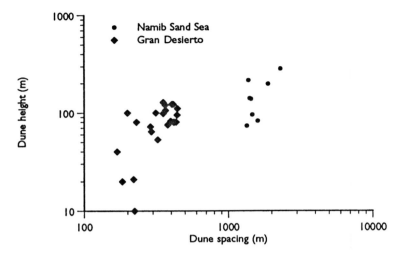

Figure 3.15 Relations between star dune height and spacing.

the subsidiary arms, their orientation depending on the winds at the time. The lower slopes of the star dunes form a wide, undulating plinth with slope angles of 2–5°. In many areas, there are small crescentic and reversing dunes on the plinths. These dunes merge with the arms descending from crestal ridges. Plinths of star dunes are often quite well vegetated and there are deep hollows and blowouts between and adjacent to the arms descending from the crestal areas. Interdune areas between star dunes are often irregular in shape, with areas of small crescentic and reversing dunes in some places.

Star dunes are a prominent feature of the Gran Desierto Sand Sea and occupy some 10 per cent of its area south and southwest of the Sierra del Rosario (Lancaster *et al.* 1987). Many of the star dunes occur in linear clusters or chains with WNW–ESE trends (Plate 11b). The clusters are mostly 2–5 km long and 1–2 km wide with a spacing of 2–3 km. Each star dune cluster or chain consists of a series of 80–100 m high, sharp crested, straight to slightly sinuous, near symmetrical ridges on a dominant NE–SW alignment and with a spacing of 300–400 m. The avalanche face of these ridges is oriented to the NW or NNW in summer, but is reversed seasonally to face southwest or south in the winter. Lower linear crests on NW–SE or N–S alignments form subsidiary arms on many star dunes, and may connect adjacent NE–SW trending ridges to form a rectilinear pattern of peaks and ridges separated by deep hollows, in which pre-dune alluvial deposits are often exposed. The main NE–SW trending star dune arms lie on a wide plinth, on the lower parts of which are multiple transverse and reversing ridges 3–10 m high on ENE–WSW or NE–SW alignments. These dunes increase in height to 10 –20 m and become more symmetrical in cross section towards the main dune ridges. The interdune corridors between the star dunes consist of areas of 2–3 m high partly vegetated crescentic dunes and sand sheets.

Wind regimes of star dunes

Dunes of star form are associated with wind regimes that are multi-directional, or complex, especially in the months in which most sand transport occurs (Aufrère 1928; McKee 1966; Fryberger 1979). Annual resultant or net sand transport in these situations is often low. Examples of wind regimes in areas of star dunes are given in Figure 3.16.

Fryberger (1979) showed that star dunes occur in areas which exhibit a wide range of annual potential sand transport rates, but are characterized by low to very low ratios between total and net potential sand transport with

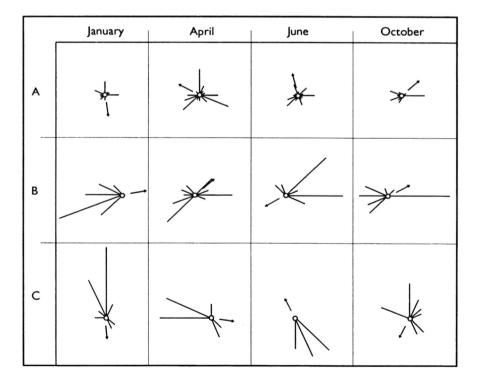

Figure 3.16 Wind regimes of star dunes (from Lancaster 1989d). A: Ghudamis, Grand Erg Oriental (after Fryberger 1979); B: Eastern Namib Sand Sea; C: Gran Desierto Sand Sea.

annual unidirectional indices from 0.20 to 0.40. At Ghudamis, on the margin of the area of star dunes in the Grand Erg Oriental, there is a complex wind regime in all months, but two major directional sectors can be distinguished: NNE–E in summer and W–SW in winter (Figure 3.16a). Wind regimes of areas where star dunes occur in the Namib Sand Sea are characterized by a tri-modal wind regime, with SW–W and NE–E modes dominating (Lancaster 1985a) (Figure 3.16b). Although no wind records exist for the area of the Gran Desierto, data from Yuma, Arizona, supported by observations in the field, suggest that winds in the region are from three major sectors, the importance of which varies seasonally (Figure 3.16c). Winter northerly, spring westerly and summer southerly winds each generate 25–30 per cent of annual potential sand transport.

Complexity in wind regimes in desert areas appears to be the result of circulation patterns which vary significantly from season to season, combined with the effect of the passage of frontal systems during winter and spring. A situation in which a winter high pressure cell is replaced by a thermal low in summer occurs in the Sonoran and Mojave Deserts of North America, and in central Asia. In the northern Sahara, wind regimes change as winter westerlies associated with the passage of frontal systems through the Mediterranean are replaced in summer by anticyclonic circulations and northerly to northeasterly winds (Dubief 1952). Many areas of star dunes are associated with the poleward margins of desert regions, where the effects of seasonal changes in wind directions are most often felt, compared to the equatorial margins where wind patterns are dominated by trade-wind circulations.

Parabolic dunes

Parabolic dunes (Plate 12), common in many coastal and semi-arid dune fields, have a restricted distribution in arid region sand seas. The only major sand sea with significant areas of this dune type is in the Thar Desert of India (Verstappen 1968; Breed et al. 1979; Wasson et al. 1983). Areas of parabolic dunes occur in the southwestern Kalahari (Thomas and Shaw 1991b), Saudi Arabia (Anton and Vincent 1986), northeast Arizona (Hack 1941), in eastern Colorado (Muhs 1985), the Nebraska Sand Hills (Ahlbrandt and Fryberger 1980; Ahlbrandt et al. 1983), and at White Sands (McKee 1966).

Parabolic dunes are characterized by a U shape with trailing partly vegetated parallel arms 1–2 km long, and an unvegetated active 'nose' or dune front 10–70 m high that advances by avalanching. In the Thar Desert, compound parabolic dunes with shared arms occur in some areas and result from merging or shingling of several generations of dunes with different migration rates (Wasson et al. 1983). The conditions under which parabolic dunes form are not well known. They seem to be associated with the presence of a moderately developed vegetation cover, and with unidirectional

Plate 12 Parabolic dunes on the Snake River Plains, Idaho (photograph courtesy of R. Greeley).

wind regimes. Downwind, some parabolic dunes are transitional to crescentic dunes (Anton and Vincent 1986).

Nebkhas

Nebkhas (also known as coppice dunes) develop where sand is trapped by vegetation clumps (Thomas and Tsoar 1990). They are widely distributed in semi-arid areas, and also occur in hyper-arid areas like the coastal Namib, where strong sand movement into areas of phreatophyte vegetation occurs (Lancaster 1989b). Nebkhas in the Namib (Plate 13) range up to 3.5 m high. In Mali, they are 0.35 to 0.72 m in height (Nickling and Wolfe 1994) and have developed in the last thirty years.

Lunettes

Dunes of a 'U' or sub-parabolic shape may form on the downwind margins of small playas. Such dunes are widely distributed in Australia (Bowler 1983) and the Kalahari (Lancaster 1978; Goudie and Thomas 1985). In some places, linear dunes also extend downwind from lunettes in Australia (Twidale 1980) and the southwestern Kalahari (Lancaster 1988b) (Figure 3.17). Like nebkhas, wind regime is not a primary factor influencing the form and development of lunettes.

Plate 13 Nebkha anchored by !nara in the Kuiseb delta region, Namibia.

Zibars and sand sheets

Not all aeolian sand accumulations are characterized by dunes. Low relief sand surfaces such as sand sheets are common in many sand seas, and occupy from as little 5 per cent of the area of the Namib Sand Sea to as much as 70 per cent of the area of Gran Desierto (Lancaster *et al.* 1987). Fryberger and Goudie (1981) estimated that 38 per cent of aeolian deposits are of this type. Many sand sheets and interdune areas between linear and star dunes are organized into low rolling dunes without slipfaces, known as zibars (Holm 1960; Warren 1972; Tsoar 1983a; Nielson and Kocurek 1986) with a spacing of 50–400 m and a maximum relief of 10 m. Typically, zibars (Plate 14) are composed of coarse sand, and occur on the upwind margins of sand seas. Zibars in the Skeleton Coast dune field are 1–2 m high with a crest to crest spacing of 100 m (Lancaster 1982a). Those on the western margins of the Algodones dune field have a crest-to-crest spacing of 60 m and rarely exceed 2 m in height. Somewhat similar coarse grained bedforms occur in the Selima Sand Sheet of Egypt, but have a much longer wavelength (130–1200 m) and an amplitude of 0.1–10 m (Breed *et al.* 1987; Maxwell and Haynes 1989).

Sand sheets develop in conditions unfavourable to dune formation (Kocurek and Nielson 1986). These may include a high water table, periodic

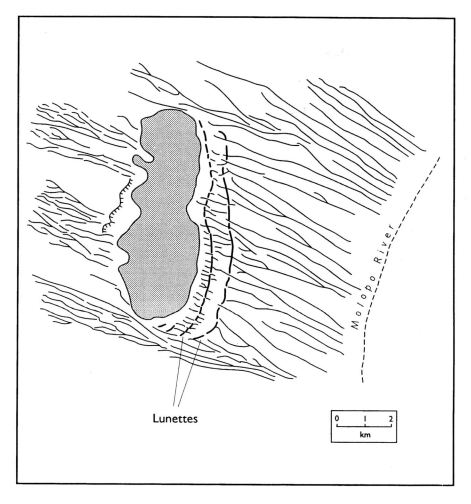

Lunettes

Figure 3.17 Lunette dunes and linear dunes in the southwestern Kalahari (from Lancaster 1988b).

flooding, surface cementation, coarse grained sands (Plate 15a), and presence of a vegetation cover that act to limit sand supply for dune building (Plate 15b). Sand sheets that develop under different conditions appear to have distinct geomorphic characteristics. Some of the most extensive sand sheets known are located in the eastern Sahara, where they cover more than 100,000 km². Sand sheets in this area have a relief of less than 1 m over wide areas and a total thickness of a few cm to as much as 10 m (Breed *et al.* 1987).

Plate 14 Zibars between linear and crescentic dunes, Algodones dune field, California.

The sand sheets consist of sandy plains with a granule to pebble lag deposit that forms the surface layer, together with areas of very low amplitude (10–30 cm), long wavelength (130–1200 m) bedforms that form a giant chevron pattern on Landsat images (Maxwell and Haynes 1989). In this area, sand sheets develop because of an abundance of coarse particles that inhibit all but localized dune development.

Sand sheets cover an area of 1000–1500 km² in the northwestern parts of the Gran Desierto. They form a sparsely vegetated, flat to gently undulating surface with a maximum local relief of 1–5 m (Plate 15b). The sand sheets are composite features consisting of successive generations of aeolian accumulations separated by stabilization surfaces (Lancaster 1993a). Sand sheets in the northwestern Gran Desierto appear to have developed in conditions of a restricted sand supply and a sparse vegetation cover which is insufficient to prevent sand transport taking place, but sufficient to cause divergence and convergence of airflow around individual plants in the manner suggested by Ash and Wasson (1983) and Fryberger *et al.* (1979), giving rise to localized deposition by wind ripples and shadow dunes.

(a)

(b)

Plate 15 (a) Sand sheet developed in coarse, poorly sorted sand,
Elizabeth Bay, Namibia.
(b) Sparsely vegetated sand sheet in the northern Gran Desierto Sand Sea.

(a)

(b)

Plate 16 Climbing and falling dunes.
(a) Climbing dunes in the Mojave Desert.
(b) Falling dune: the Cat Dune, Mojave Desert.

Topographically controlled dunes

Dunes that owe their existence and form to interactions of sand-transporting winds and topographic obstacles are common in many desert regions, yet are poorly documented. At the smallest scale are shadow dunes that accumulate in the lee of large boulders or breaks in slope. By contrast, large climbing and falling dunes in the Mojave Desert have a total relief of as much as 200 m and comprise large volumes of sand (Evans 1961; Tchakerian 1992). Very large climbing dunes also occur in the Atacama Desert of Peru (Howard 1985).

Echo dunes form on the windward side of escarpments, where the average slope is greater than 55° and the dune is separated from the scarp by a sand-free area that results from a fixed eddy between the cliff and the dune (Tsoar 1983b). When average slope angles are less than 55°, the fixed eddy is small or non-existent and the result is a climbing dune (Plate 16a). Falling dunes (Plate 16b) occur in the lee of hills that are asymmetric in form, with lee slopes steeper than those to windward. In many areas of the Mojave Desert, climbing and falling dunes occur on opposite sides of the same mountain mass (Smith 1954).

In some areas, linear dunes form in the lee of mountains or downwind from escarpments (Clos-Arceduc 1967; Smith 1978; Breed et al. 1979; Tsoar 1989) as the wind field is modified by the topographic obstacle. Linear dunes extend for 1–2 km downwind from the Superstition Mountains of southern California, but break down into a series of small barchans thereafter.

CONCLUSIONS

Although field studies indicate that there is a wide variety of different dune morphological types, remote sensing images show a remarkable degree of similarity between different sand seas and dune fields, suggesting that there are some overall controls of dune morphology. Five major dune types can be identified: crescentic, linear, star, reversing, and parabolic. Each major dune type can occur in three varieties: simple (individual dunes); compound (superimposed dunes of the same type) and complex (merging or super-imposition of dunes of different types). Together with wind ripples, simple and compound or complex dunes form a hierarchical system of aeolian bedforms. Dune morphology and occurrence appears to be controlled primarily by regional wind regime characteristics (especially directional variability), but is also influenced by vegetation cover and sediment availability. The morphology of other dune types is not primarily determined by wind regime characteristics. Sand sheet and zibar development is mainly determined by the availability of coarse sand, but a sparse vegetation cover and sediment starvation may be locally important. The formation of parabolic dunes and nebkhas is strongly influenced by interactions between sand transport and vegetation cover, whereas climbing and falling dunes are anchored by topographic obstacles.

In all areas, there are close correlations between dune height and spacing, suggesting some overall self-organization of the system that is related to sediment availability which sets a limit to dune size and the pattern of airflow over and between the dunes, which may determine their spacing. Fundamentally, dune morphology is the product of interactions between sand transporting winds and the dune, which will be examined in Chapter 5. The controls of dune morphology and morphometry will be discussed in detail in Chapter 6.

4

DUNE SEDIMENTS

INTRODUCTION

Dunes are depositional landforms. As a result their sediments provide a record of dune accumulation and can therefore provide important information on the processes of dune formation and development.

There are three primary modes of deposition on dunes (Hunter 1977): (1) migration of wind ripples, (2) fallout from temporary suspension of previously saltating grains in the flow separation zone in the lee of the crest, and (3) avalanching on the lee slope of the dune. These processes form three main types of aeolian sedimentary structures which constitute the primary units of aeolian deposition (Hunter 1977; Kocurek and Dott 1981): (1) climbing translatent strata (wind ripple laminae), (2) grainfall laminae, and (3) grainflow cross-strata.

Climbing translatent strata are formed by the migration of wind ripples under conditions of net deposition that give rise to bedform climbing (Rubin and Hunter 1982). In conditions where wind shear stress decreases downwind, the transport capacity of the wind declines and excess sediment is deposited or transferred to the bed, producing wind ripple laminae (Plate 17a). These are equivalent to the traction or accretion deposits of Bagnold (1941) and Sharp (1966). Wind ripple deposits are very widespread in most dune areas and occur on the stoss slopes of many dunes, as well as on the plinths or aprons of larger linear and star dunes (Kocurek 1986; Fryberger and Schenk 1988). Wind ripple laminae are also prominent in the deposits of most sand sheets and interdune areas (Ahlbrandt and Fryberger 1981; Kocurek and Nielson 1986) and on the steeply sloping lee faces of dunes affected by strong secondary flows (Tsoar 1983a; Havholm and Kocurek 1988).

Grainfall laminae (Plate 17b) are formed when grains that saltate over the brink line of the dune come to rest on the lee face. Preservation of grainfall deposits only occurs if no subsequent oversteepening and avalanching take place. They appear to be more common on small dunes or in deposits laid down in very strong winds so that grainfall occurs on the lower parts of the lee face (Hunter 1977).

(a) Wind ripple laminae on linear dune plinth.

(b) Grainflow cross strata.

(c) Grainfall laminae and grainflows, Great Sand Dunes, Colorado.

Plate 17 Primary sedimentary structures in dunes

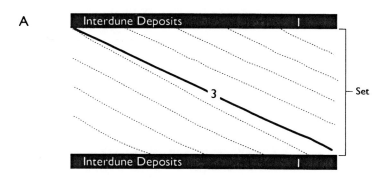

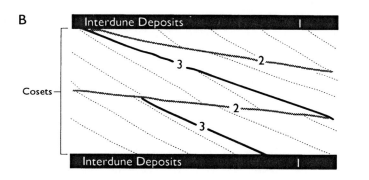

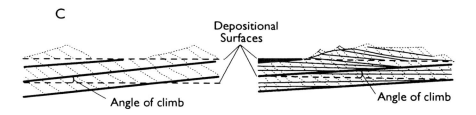

Figure 4.1 Model for the formation of bounding surfaces in aeolian deposits (after Kocurek 1988). A: migration of simple dunes and interdune areas; B: migration of compound/complex dunes and interdune areas. Different orders of bounding surfaces indicated by 1, 2, 3.

Grainflow cross-strata (Plate 17c) are formed by avalanching of grainfall deposits on the lee face of dunes which are oversteepened beyond the angle of repose of dry sand (28–34°) (Anderson 1988). They consist of a series of overlapping tongues 3–4 cm thick with coarse grains concentrated on their upper surfaces and at the toe of the tongue. Most grainflow strata thin out towards the top of the dune, and have a tangential lower contact with the base of the dune. Slump structures and blocks may be preserved in damp sand avalanches (McKee *et al.* 1971).

Primary dune sedimentary structures are separated by bounding surfaces of different types that can be classified in terms of a hierarchical scheme (Figure 4.1) (Brookfield 1977; Fryberger 1991). Third order or primary bounding surfaces occur within sets of laminae and represent reactivation episodes resulting from short-term changes in wind strength and/or direction; second order or growth surfaces bound sets of strata and form by erosion or non-deposition as dunes grow episodically; first order or stacking surfaces may divide the accumulations of laterally migrating dunes, or in some environments may represent episodes of deflation to the water table (Stokes 1968; Loope 1984). In addition, regional scale or super surfaces (Kocurek 1988) form as a result of hiatuses in sand sea accumulation due to depletion of sediment supply and/or climatic changes. Their characteristics and significance are discussed in Chapters 7 and 8.

DUNE SANDS

Composition and sources

Dune sediments are dominated by quartz and feldspar of sand size (63 to 1000 μm), derived originally from the weathering of quartz-rich rocks, especially granites and sandstone. Dune fields close to source regions (e.g. Kelso dunes in the Mojave Desert) often contain large proportions of feldspar and lithic fragments. In addition, dunes composed of volcaniclastic materials (e.g. pumice, cinders, basalt, tuff) occur in some areas (Edgett and Lancaster 1993). Examples are to be found at the Great Sand Dunes, Colorado (51.7 per cent volcanic rock fragments, 27.8 per cent quartz), in central Nevada, and southeastern Oregon. Inland dunes of carbonate sand have been documented from Oman and the United Arab Emirates (Besler 1982; Gardner 1988), and gypsum dunes occur at White Sands (McKee 1966); in the eastern Great Basin and adjacent to the chotts of Tunisia (Besler 1977; Mainguet and Jacqueminet 1984).

Direct contributions of sand from the weathering of bedrock appear to be limited. Sandstones, some of them aeolian in origin, have been cited as sources of sand for dunes in the Namib (Besler and Marker 1979), the Thar Desert (Wasson *et al.* 1983), Saudi Arabia (Holm 1960) and north America (Ahlbrandt 1974a). In most cases, however, aeolian sand is derived from materials that have been transported by some other medium. The most important sediment sources are fluvial and locally deltaic sediments (e.g. Glennie 1970; Andrews 1981; Lancaster and Ollier 1983; Wasson 1983; Blount and Lancaster 1990). Other sources include beaches (e.g, Inman *et al.* 1966; Lancaster 1982a; Allison 1988; Corbett 1993), Pleistocene paleolakes (McCoy *et al.* 1967) and playas and sebkhas (e.g. McKee 1966; Besler 1982).

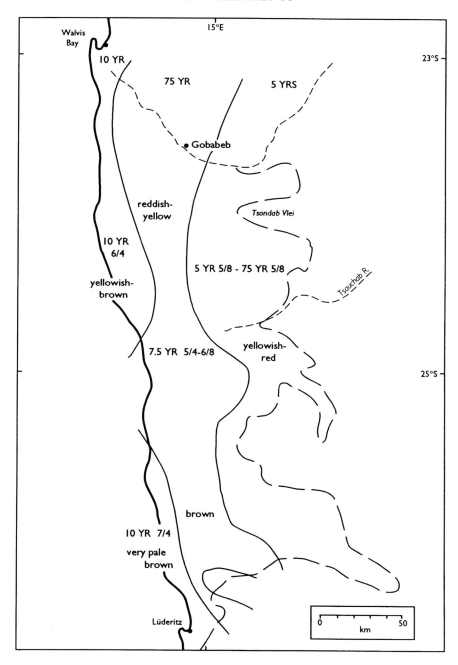

Figure 4.2 Namib sand colours (after Lancaster 1989b).

Many sand seas have internal sediment sources as a result of the deflation of interdunes and reactivation of older dunes (Folk 1971; Wasson 1983; Lancaster 1988b; Kocurek *et al.* 1991). In the Sahara and Arabia, some sand seas receive inputs of sand from adjacent sand bodies along well-defined sediment transport corridors that are clearly visible on satellite images (Fryberger and Ahlbrandt 1979; Mainguet 1984b).

Tracing dune sands to their sources involves comparison of their mineralogy and chemical composition with that of potential source areas. This can be done using analyses of both light and/or heavy minerals (e.g. Sharp 1966; Merriam 1969; Wiegand 1977; Lancaster and Ollier 1983). Recently, minor element geochemistry has proved a valuable technique and ratios of the elements Rb to Sr and Ti to Zr can effectively pin-point sand sources (Muhs in press).

Sand colours

The colours of dune sands vary widely from one sand sea to another, and even within the same sand sea (Figure 4.2). Some of the darkest and most reddened sand is in the Kalahari and Simpson-Strzelecki deserts where sand colours are red to yellowish red (Munsell Notation 2.5YR 5/8 to 7.5YR 5/8) (Folk 1976a; Wasson 1983; Lancaster 1986). Relatively pale colours (10YR 6/2 to 7/4) are associated with dunes in some coastal deserts (e.g. Lancaster 1982a). Most North American sand seas and dune fields are composed of relatively pale coloured sand (10YR 6/2) (e.g. Blount and Lancaster 1990; Lancaster 1993b). Sand from the northwestern Sahara (Alimen *et al.* 1958); Libya (Walker 1979); the central Namib Sand Sea (Lancaster 1989b) and Arabia (Besler 1982) is intermediate in colour with reddish yellow (7.5YR 5/6 to 5/8) to yellowish red (5YR 5/8) colours common.

Studies of individual sand grains under the microscope show that increased reddening of the grains is achieved by a greater extent and thickness of iron oxides deposited in pits and other surface irregularities on clear or frosted quartz grains. Reddening is most pronounced on smaller (125–250 μm), more angular grains, as was observed by Folk (1976a) from Simpson desert dunes. In some areas, red colours are the result of the amber colour of many quartz grains (e.g. Wasson 1983).

It has been widely reported that dunes become redder with distance in the direction of transport and hence as the sand in them becomes older (e.g. Alimen *et al.* 1958; Logan 1960; Wopfner and Twidale 1967; Folk 1976a; Breed and Breed 1979; Walker 1979; El Baz 1978). However, Gardner and Pye (1981) suggest that colour is not necessarily a function of dune age and that time is not always an independent variable. As observed by Folk (1976a), reddening is a product of moisture availability, temperature and time. Also important are grain mineralogy, presence of weatherable minerals and the aeolian dust input. Further, different sources for sands, especially the existence of pre-reddened sediments, may be important (Wasson 1983).

Grain morphology

Although early workers (e.g. Shotton 1937; Cailleux 1952) suggested that aeolian sands were rounded or well rounded in shape, more recent investigations (Folk 1978; Goudie and Watson 1981; Goudie *et al.* 1987) indicate that, in aeolian sands, true roundness in the dominant 125–250 µm size group is rare and most grains are sub-angular to sub-rounded in shape. Goudie and Watson (1981) also noted that grains from different sand seas cluster around distinctive grain roundness characteristics that reflect sand source characteristics and transport pathways. Thus, sand from Tunisian dunes is more rounded than that from the Namib and Kalahari, which in turn is more rounded than Thar and some North American dune sand, much of which has been derived directly from Mesozoic and later source rocks (Figure 4.3).

The surfaces of desert sand grains may have distinctive characteristics when examined at high magnifications using a scanning electron microscope (Krinsley and Trusty 1985). Some of these features are: (1) rounding of edges on all grain shapes, (2) 'upturned plates' resulting from breakage of quartz along cleavage planes in the crystal lattice, (3) elongate depressions resulting from conchoidal fractures during collisions between grains; (4) smooth

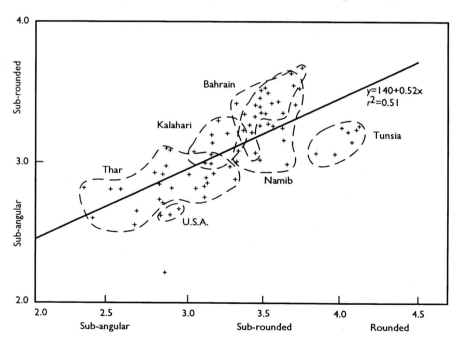

Figure 4.3 Variation in grain shape between sand seas (after Goudie and Watson 1981).

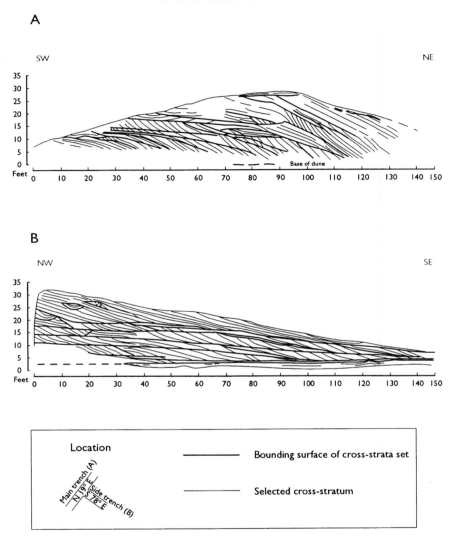

Figure 4.4 Structures in barchan dunes at White Sands (modified from McKee 1966). A: parallel to wind direction; B: perpendicular to wind direction.

surfaces resulting from solution and re-precipitation of silica, and (5) arcuate, circular or polygonal fractures resulting from collisions and/or weathering. The proportions of these features change through time in accumulating aeolian deposits so that older sand grains exhibit a greater proportion of features that can be attributed to weathering (Tchakerian 1992).

SEDIMENTARY STRUCTURES OF MAJOR DUNE TYPES

Crescentic dunes

Studies by McKee (1966, 1979b), Ahlbrandt (1974b) and Hunter (1977) have shown that the sediments of barchans and crescentic ridges (Figures 4.4, 4.5) are dominated by grainfall and grainflow cross-strata deposited on the lee face and preserved as the dune migrates downwind. Wind ripple laminae form a thin set of strata that parallel the slope of the stoss slope and dune crest. Barchans in northern Arizona, at White Sands and in the Kilpecker dune field are composed of sets of cross-strata with dips between 31 and 34° in the direction of dune migration. The sets of cross-strata are separated by horizontal to steeply-dipping bounding surfaces. Crescentic ridges at Kilpecker, White Sands, and in the Namib Sand Sea are dominated by large sets of leeward-dipping avalanche face laminae at angles of 30 to 34°, with a tangential contact to the lower surface. Wind ripple laminae are restricted in all cases to a thin veneer over the stoss slope of these dunes. Detailed studies of small barchanoid ridges at Padre Island by Hunter show the spatial distribution of grainfall, grainflow and wind ripple laminae (Figure 4.6). Larger dunes are dominated by grainflow laminae, whereas small dunes include more grainfall and wind ripple laminae.

Linear dunes

Linear dunes are composed of varying proportions of wind ripple and avalanche face laminae. Bagnold (1941) published a hypothetical section of a linear dune, in which laminae were divided into steeply dipping avalanche deposits on the crest and central areas of the dunes, and low angle accretion laminae on the dune plinths (Figure 4.7). McKee and Tibbitts (1964) confirmed this basic pattern for a 15 m high simple linear dune in the Libyan desert (Figure 4.8a). They found that the upper parts of the dunes were composed of cross-strata with dips of 26–34°, with low angle strata (4–14°) in lower areas. The high angle strata were interpreted as avalanche face deposits formed in a diurnally bi-directional wind regime, with winds blowing at around 45° to the dune. By contrast, internal structures of simple linear dunes in the Simpson Desert and northeast Arizona studied by Breed and Breed (1979) consisted mostly of medium-scale thin cross beds with dips commonly less than 20° occurring in tabular and wedge shaped sets separated by near horizontal bounding surfaces (Figure 4.8b).

Tsoar (1982) recognized two groups of laminae on a simple linear dune in the Sinai (Figure 4.9). The first group were deposited by grainfall and grainflow in a 1–2 m wide area parallel to the crest and dipped at 33° perpendicular to the crest line, whereas the second, with dips of 20–25°

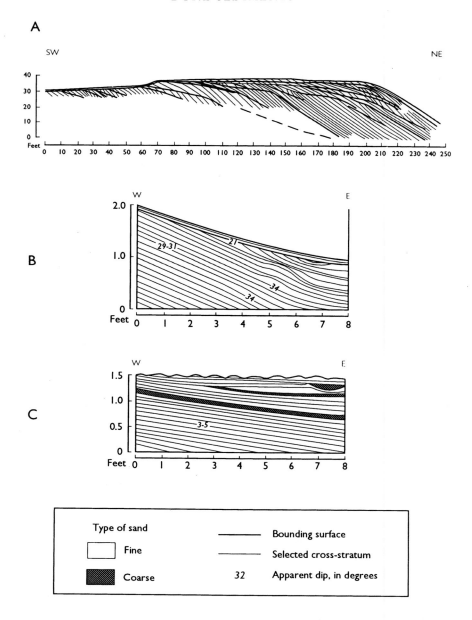

Figure 4.5 Structures in crescentic dunes (modified from McKee 1979b). A: White Sands, New Mexico, parallel to wind direction; B, C: Kilpecker Dunes, Wyoming, parallel to dominant wind on stoss slope (B) and lee slope (C).

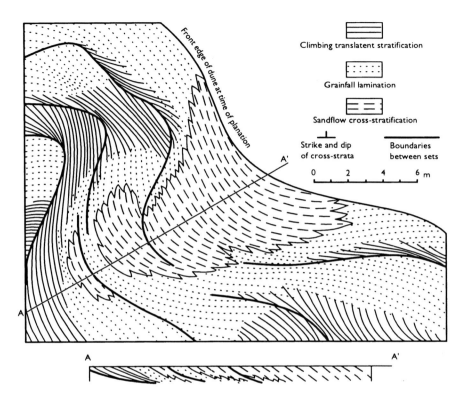

Figure 4.6 Spatial distribution of sedimentary structures in a small crescentic ridge (after Hunter 1977).

oblique to the crest line, were deposited by accretion of wind ripple deposits on pre-existing slopes as a result of local changes in flow velocity. These deposits formed the bulk of the dune sediments. Avalanche and wind ripple deposits accumulate on each side of the dune according to the primary wind direction in each season. Very similar sequences of deposits were recognized on simple linear dunes in the southern Namib by Lancaster (1989b).

Structures of large complex linear dunes were studied by McKee (1982) from several areas on the north margin of the Namib Sand Sea. In crestal areas of the dunes, avalanche face strata dip at 28–35° towards the interdune on both sides of the dune crest. On the upper parts of the plinths, deposits consist of wind ripple laminae with dip angles of 10–20°, declining to 8° or less towards the interdunes. Complex linear dunes in Mauritania and Saudi Arabia (McClure 1978; Kocurek *et al.* 1991), and probably elsewhere, are however composite features that consist of deposits that have accumulated at

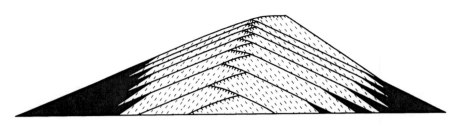

Figure 4.7 Model of linear dune structure (modified from Bagnold 1941).

several different periods (see Chapter 8). The bulk of sediments in these dunes were laid down in the late Pleistocene and only their upper parts exhibit sedimentary structures formed in contemporary conditions.

It is probable that the variations in sedimentary structures observed in linear dunes in different sand seas are a result of different wind regime characteristics, as suggested by Lancaster (1982b), and reflect varying proportions of avalanche versus wind ripple deposits. Dunes in areas where formative winds blow at a small angle to the dune, as in Sinai, the southern Namib, and Australia, are likely to be characterized by a higher percentage of wind ripple laminae deposited by along-slope secondary flows and less grainfall and grainflow laminae compared to dunes in areas where winds blow at a high angle to the dune, as in the northern Namib Sand Sea and Libya, and give rise to significant avalanche face deposition without reworking by secondary winds.

Star dunes

The sedimentary structures developed in star dune deposits are complex, although investigations by trenching only sample a very small proportion of the total accumulation. McKee (1966; 1982) trenched star dunes in Saudi Arabia and the Namib Sand Sea and observed cross-cutting sets of high angle cross-strata near dune crests (Figure 4.10). These deposits represent avalanche faces formed by winds from multiple directions. Structures in the crestal areas of star dunes in the Namib Sand Sea (McKee 1982) also show clear evidence of seasonal wind reversal. Avalanche face grainfall and grainflow laminae dipping east at 35° were overlain by another set which covers their eroded tops and dips towards the west at 33°. Similar overlapping sets of avalanche face laminae were observed in reversing dunes at Kelso and Great Sand Dunes (Sharp 1966; McKee 1979b). Complex structures were also intersected on star dune arms in both examples investigated. McKee also observed that, in basal areas of star dunes in the

A

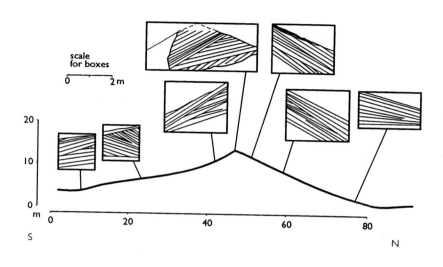

B

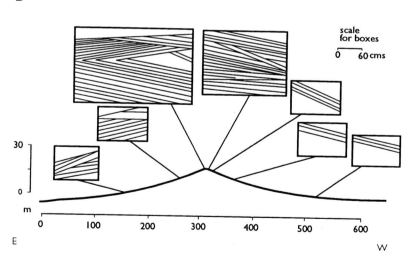

Figure 4.8 Linear dune structures. A: modified from McKee and Tibitts 1964; B: modified from Breed and Breed 1979.

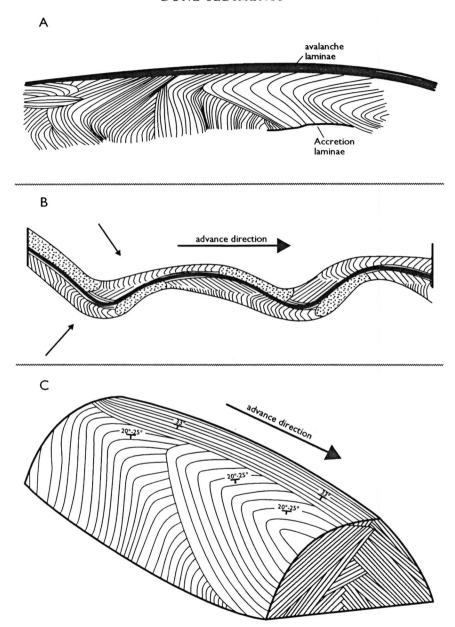

Figure 4.9 Linear dune structures (after Tsoar 1982). A: view perpendicular to dune; B: areas of erosion and deposition; and C: model for dune structures based on field observations.

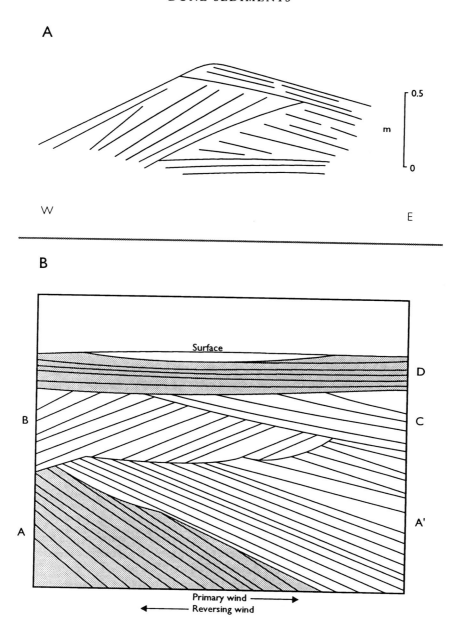

Figure 4.10 A: Star dune structures, Namibia (drawn from photograph in McKee 1982); B: Reversing dune structures at Great Sand Dunes (after McKee 1979b).

Namib, most avalanche face laminae dip steeply to the northeast with low angle strata dipping westward. This pattern was interpreted to mean that the star dune accumulated largely by winds from south to west and east to southeast directions. Similar asymmetry in the deposits of reversing dunes has been noted from Great Sand Dunes by Andrews (1981), indicating that deposits developed by secondary winds are largely reworked by those from the primary wind direction. Recent ground-penetrating radar studies of dunes at Great Sand Dunes indicate a complex series of trough-shaped bounding surfaces formed by scour in a reversing wind regime and/or by secondary lee side airflow (Schenk *et al.* 1993).

Nielson and Kocurek (1987) documented the widespread occurrence of low-angle wind ripple deposits on star dunes at Dumont, California, where dune plinths form most of the depositional surfaces. Because of their low topographic position, such deposits have a higher potential for preservation in the rock record than the avalanche face deposits which are confined to a small part of the crestal areas of the dune. Kocurek (1986) and Nielson and Kocurek (1987) argue that star dunes therefore may not leave a characteristic suite of deposits, which could explain why star dunes have rarely been recognized in the rock record (Clemmensen 1987).

Parabolic dunes

Parabolic dune structures described by McKee (1966) from White Sands and Ahlbrandt (1974b) from Wyoming (McKee 1979b) (Figure 4.11) exhibit sets of cross-strata that dip downwind at angles of 22–34°. On the stoss slopes and dune arms deposits are typically wind ripple laminae in 1–2 m thick sets. Set boundaries tend to be convex downwind, reflecting undercutting of the dune nose by cross-winds according to McKee, or low rates of deposition parallel to existing dune slopes. Bioturbation of dune structures is also common, reflecting the importance of vegetation in the development of these dunes.

Zibars and sand sheets

Sediments of sand sheets and zibars are dominated by wind ripple laminae, reflecting the dominant surface process in these environments. Deposits of zibars at the Algodones Dunes documented by Nielson and Kocurek (1986) consist entirely of lee face wind ripple laminae deposited as the zibar migrates downwind. The wind ripple deposits consist of fine and medium sand laminae, with occasional coarse sand laminae resulting from high wind events, that alternate as both packages up to 10 cm in thickness and on a laminae-by-laminae basis (Figure 4.12). Laminae dip at less than 15° and packages of laminae are bounded by low angle truncation surfaces resulting from changes in wind direction. Interzibar deposits are thin (< 20 cm) and

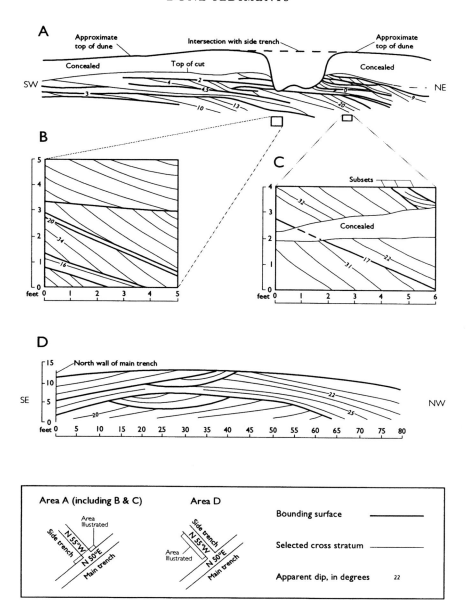

Figure 4.11 Parabolic dune structures at White Sands (after McKee 1966). A, B, C: parallel to wind direction; D: perpendicular to dominant wind.

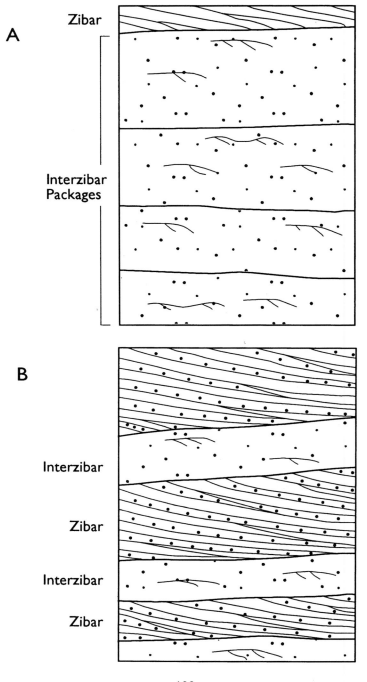

Plate 18 Sand sheet structures in the northwestern Gran Desierto.

consist of coarser grained and lower angle (< 5°) wind ripple laminae with numerous truncation surfaces.

Sand sheet deposits are very similar to those of zibars and consist of planar sets of horizontal to very gently dipping (2 to 5°) wind ripple laminae (Fryberger *et al*. 1979; Kocurek and Nielson 1986; Lancaster 1993a). In some areas, such as Great Sand Dunes, and the northern margins of the Gran Desierto sand sea, sand sheet deposits consist of alternating 1–5 cm thick sets of coarse grained and fine grained laminae (Plate 18). Elsewhere, structureless sands may occur alone or in combination with laminated units. In some instances structureless sands are the result of bioturbation, pedogenesis, or root growth (e.g. Fryberger *et al*. 1979; Lancaster 1993a); but they may also result from deposition in conditions of moderate vegetation cover (Kocurek and Nielson 1986). Where coarse sands are available, as in the Western Desert of Egypt, many sand sheets develop a surface armour of coarse sand or granules that form granule ripples (Breed *et al*. 1987; Maxwell and Haynes 1989; Fryberger *et al*. 1992). Many sand sheets are vegetated to some extent

Figure 4.12 Sedimentary sequences in zibar, based on observations at the Algodones Dunes (after Nielson and Kocurek 1986). A: amalgamated interzibar deposits that lack distinctive sedimentary structures or textures; B: zibar/interzibar deposits. Note zibar wind ripple laminae with dips of less than 15°.
Interzibar laminae dip at 5° or less.

and possess 10–20 cm of micro relief, resulting in an irregular pattern of deposition that is characterized by small-scale scour and fill structures, reactivation and truncation surfaces, convex upward laminae, bioturbation, and root traces (Fryberger *et al.* 1979). In other areas, especially where sand sheets are developed on the margins of dune fields, they may include non-aeolian sediments, such as mud drapes, distal fluvial deposits, or evaporites. Where the water table is high, sand sheets may be transitional to siliciclastic sebkhas (Fryberger *et al.* 1983).

GRAIN SIZE AND SORTING CHARACTERISTICS OF DUNE SANDS

The characteristics of the primary modes of aeolian deposition are reflected in the size and sorting characteristics of aeolian sands. Most dunes are composed of fine to medium (mean grain size 1.60–2.65 phi, 160–330 µm), very well to moderately sorted sands (phi standard deviations 0.26–0.55), although there is a wide range between and within individual dunes and sand seas (Ahlbrandt 1979; Lancaster 1989b). Wind ripple deposits tend to be relatively coarser and less well sorted because they contain grains transported by both reptation and saltation. Grainfall deposits are the finest and best sorted and are composed only of saltating grains. Grainflow deposits are somewhat coarser than grainflow and contain both saltating and reptating grains that have been reworked by avalanching, which brings coarse grains to the surface by dispersive stress (Bagnold 1954). They may then roll downslope to accumulate at the toe of the avalanche face. Variations in the spatial distribution of these modes of aeolian deposition over dunes, together with changes in the effectiveness of saltation and reptation processes on sloping surfaces, give rise to characteristic patterns of grain size and sorting parameters over dunes that reflect the action of sand transport processes on dunes.

Grain size and sorting parameters

Considerable attention has been paid to the development and interpretation of statistical parameters of the size and sorting of sand-sized sediments. The graphical parameters devised by Folk and Ward (1957) are widely used, as are various moment statistics (Friedman 1962). Both graphical and moment measures characterize sediment size with the modal or mean grain size and sorting using the standard deviation of grain sizes. The asymmetry of the distribution is given by its skewness: positive skewness indicates a 'tail' of fine grains, whereas negatively skewed sands have a coarse 'tail' (Table 4.1). The phi scale (Equation 4.1) is commonly used to designate the size and sorting of sand.

Table 4.1 Terminology of graphical statistics of grain size and sorting.

Graphic standard deviation (phi sorting)		Graphic skewness (phi skewness)	
very well sorted	< 0.35	very positively skewed	+0.3 to +1.0
well sorted	0.35–0.50	positively skewed	+0.1 to +0.3
moderately well sorted	0.50–0.70	symmetrical	+0.1 to –0.1
moderately sorted	0.70–1.00	negatively skewed	–0.1 to –0.3
poorly sorted	1.00–2.00	very negatively skewed	–0.3 to –1.0
very poorly sorted	2.00–4.00		

Source: Folk and Ward 1957.

Table 4.2 Comparative grain size and sorting parameters for different dune types in various sand seas.

Location	Mean (phi units)	Standard deviation (phi units)
Crescentic dunes		
Gran Desierto	2.43	0.41
Algodones Dunes	2.46	0.42
White Sands	1.61	0.59
Skeleton Coast	2.02	0.51
Namib Sand Sea	2.20	0.55
Salton Sea	2.27	0.46
Tunisia	3.25	0.53
Kelso Dunes	1.80	0.49
Linear dunes		
SW Kalahari	2.16	0.49
Simpson Desert	2.53	0.43
Namib Sand Sea	2.44	0.37
Saudi Arabia	2.67	0.32
Mauritania	2.20	0.60
Star dunes		
Gran Desierto	2.43	0.41
Namib Sand Sea	2.29	0.29
Great Sand Dunes	2.09	0.26
Kelso	2.26	0.30

$$\phi = -\log_2 d \qquad (4.1)$$

where *d* is the grain diameter in millimetres.

Recently developed grain size and sorting parameters include use of the log-hyperbolic distribution (Barndorff-Nielsen *et al.* 1982; Barndorff-Nielsen and Christiansen 1985; Vincent 1986; Hartmann and Christiansen 1988). These measures are much better than the Folk and Ward approach at characterizing skewed grain size distributions, but the interpretation of the parameters in terms of sedimentary processes and their general applicability (especially to bimodal sands) is debated (Vincent 1988).

Comparative data for average values of mean grain size and phi standard deviation (sorting) for different dune types in various sand seas (Table 4.2) show that very fine (2.51–2.77 phi; 150–180 µm) dune sands occur in the Grand Erg Oriental, Thar, Simpson Desert, and Gran Desierto sand seas. Those from the Namib and northwestern Sahara are intermediate in size (2.05–2.12 phi; 200–240 µm). Relatively coarse sands occur in the Skeleton Coast dune field, and at Kelso, the Algodones Dunes, and on the northern margins of the Gran Desierto Sand Sea.

Grain size and sorting characteristics of different dune types

Crescentic dunes

Changes in grain size and sorting parameters over barchans and crescentic ridges show a consistent pattern (Figure 4.13) that is characterized by a decrease in modal and mean grain size from the base of the stoss slope towards the crest, with the finest sands occurring on the middle of the slip face (Barndorff-Nielsen *et al.* 1982; Vincent 1984; Watson 1986; Lancaster 1989b); sorting also improves in the same direction, although Watson (1986) found that dune crest sands were poorly sorted by comparison to other positions on the dune. In the Namib, skewness becomes more negative (fewer coarse grains) across the dune. A tendency for coarse grains to be concentrated at the base of the dune, especially on barchans, has been widely noted (e.g. Finkel 1959; Hastenrath 1967; Lancaster 1982a; Watson 1986). Unlike the case of linear and star dunes (see below), distinct sub-environments with characteristic grain size and sorting patterns are not discernible.

Linear dunes

Many investigators have noted changes in grain size and sorting over linear dunes. Bagnold (1941) observed that, in the Libyan desert, crest sands of linear dunes were finer than those from the base or plinth. Similar conclusions were arrived at by McKee and Tibbitts (1964) and Glennie (1970) for simple linear dunes in southwestern Libya and Oman; and by Alimen (1953), Lancaster (1981b) and Livingstone (1987b) for complex linear dunes in Algeria and the Namib. However, in the Simpson and Kalahari deserts, crests of simple linear dunes are coarser than their flanks (Crocker 1946; Folk 1970; Lancaster 1986). Further, some workers (e.g. Warren 1972) have reported that there are no differences in grain size and sorting over linear dunes they investigated.

Grain size and sorting patterns on simple linear dunes in the Simpson

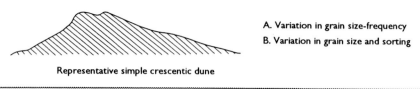

A. Variation in grain size-frequency

B. Variation in grain size and sorting

Representative simple crescentic dune

A

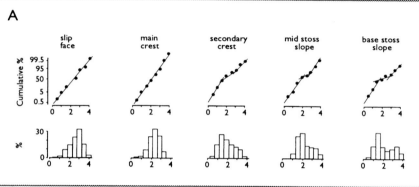

B

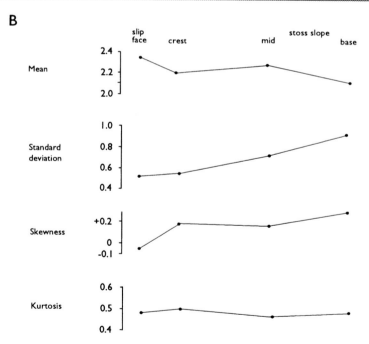

Figure 4.13 Patterns of particle size distribution (A) and grain size and sorting parameters (B) over crescentic dunes in the Namib Sand Sea.

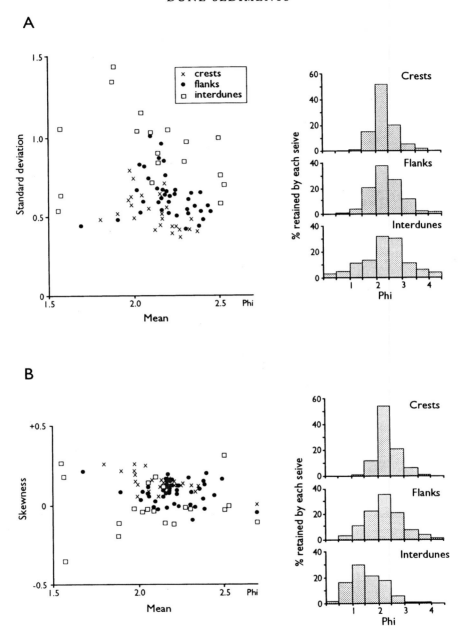

Figure 4.14 Patterns of particle size distribution and grain size and sorting parameters over simple linear dunes in the southwestern Kalahari (after Lancaster 1986).

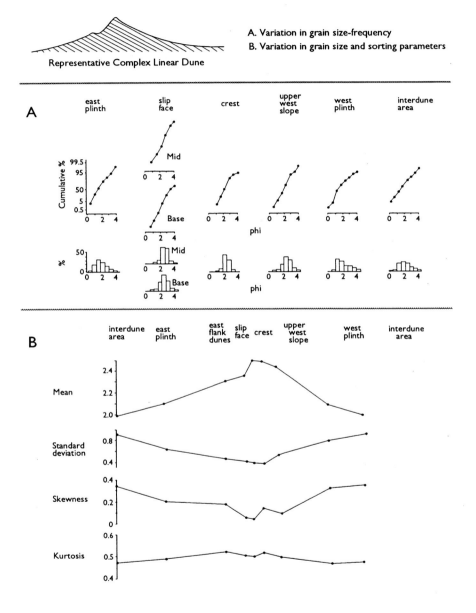

Figure 4.15 Patterns of particle size distribution (A) and grain size and sorting parameters (B) over complex linear dunes in the Namib Sand Sea.

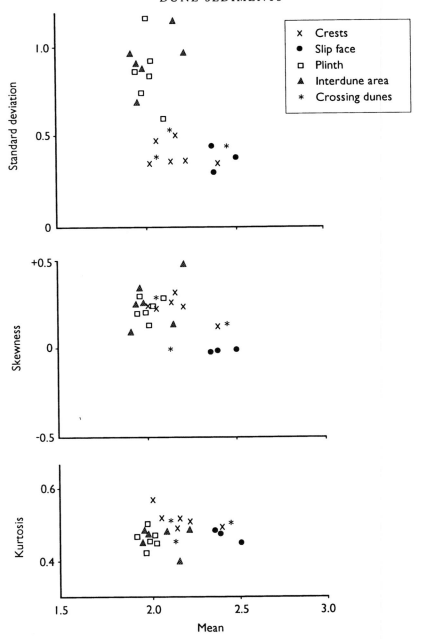

Figure 4.16 Variation in grain size and sorting from different sedimentary
sub-environments in Namib linear dune environments.

Desert studied by Folk (1970) show that the crestal sand is coarser, but better sorted, compared to flanks and interdune areas. Similar patterns were observed in the southwestern Kalahari by Lancaster (1986) (Figure 4.14). Folk explained grain size and sorting patterns on Simpson Desert linear dunes in relation to the source material. He suggested that the wind selected material in the 180 μm size range from the source sediment and concentrated it into dunes. If, as in the Simpson Desert, the source was a fine grained alluvium, then it would tend to become finer, more poorly sorted and bimodal over time as the sand was removed, leaving dune crests coarser, but better sorted than the source sediment. In the case of a coarse source material the converse would be the result. Dune flanks are intermediate in composition as they trap weakly saltating coarse sand and receive some fine sand by avalanching from slip faces.

Complex linear dunes of the Namib Sand Sea are composed of two distinct groups of sand (Figures 4.15, 4.16) (Lancaster 1981b). Sand from the crests, slip faces and upper west slopes of the linear dunes is characteristically fine, well to very well sorted and near symmetrical. By comparison, that from the plinth area are coarser; moderately sorted and frequently strongly positively or fine skewed. Interdune areas are coarser still and moderately to poorly sorted. Detailed sampling by Livingstone (1987b) indicates that the two groups of sand identified by Lancaster are end members of a continuum of grain size and sorting changes over the dune. Although Besler (1980) attributed the grain size and sorting pattern of Namib desert linear dunes to the existence of alluvial sand in basal areas and aeolian sand on the crests, the differences can be better explained in terms of the pattern of sand movement on the dunes. In this model, the population of coarse grains moved by creep and reptation are progressively left behind and concentrated in the interdune areas and on dune plinths as sand moves up dune slopes under the influence of winds from directions oblique to the dune trend (Lancaster 1981b; Livingstone 1987b).

Star dunes

As with other large dunes, the surface sand of star dunes is characterized by significant spatial variations in grain size and sorting parameters. Sand from the crests of star dunes is generally very well sorted, and often much better sorted than other dune types in a sand sea. Star dunes in the Namib Sand Sea are composed of fine to very fine sands. Mean grain size decreases from interdune areas and plinths to the crest of the dune and sorting improves in the same direction (Figure 4.17). Bivariate plots of mean values of grain size and sorting parameters show that the crests, upper slopes and slip faces of star dunes are consistently finer and better sorted than adjacent plinths and interdune areas (Figure 4.18a). Star dunes in the Gran Desierto are similarly composed of very well sorted fine to very fine sands. Sand from dune crests

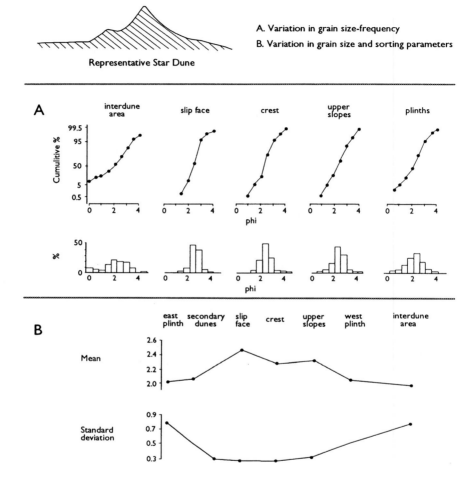

Figure 4.17 Patterns of particle size distribution (A) and grain size and sorting parameters (B) over star dunes in the Namib Sand Sea.

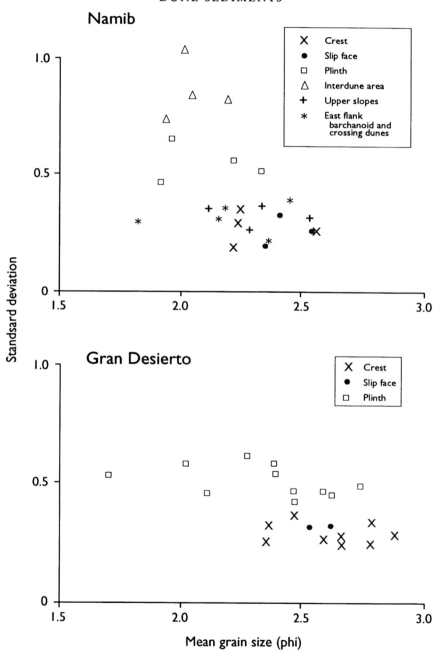

Figure 4.18 Bi-variate plots of grain size and sorting parameters for star dunes in A: Namib Sand Sea; B: Gran Desierto Sand Sea.

is typically very well sorted, and near symmetrical. Dune plinths are slightly coarser and less well sorted (Figure 4.18b). Sorting processes on star dunes are similar to those on large linear dunes, with loss of coarse grains up slope leading to their concentration in plinth areas.

Comparisons between dune types

Specific associations between dune types and grain size and sorting parameters are difficult to establish. Many of the reported relationships are probably the product of progressive sorting and fining of sands downwind from source zones. The clearest relationship to emerge is that zibar and sand sheets are composed of coarse, poorly sorted sands (e.g. Warren 1972; Kocurek and Nielson 1986). In the Algerian Sahara, Bellair (1953) found that barchans and crescentic dunes were composed of well sorted, unimodal sands, but complex linear and star dunes were composed of bi- or tri-modal sands. However, Alimen (1953) and Capot-Rey and Gremion (1964) could find no consistent relationships. At White Sands, New Mexico, McKee (1966) observed a progressive decrease in grain size and improvement in sorting from dome dunes, through barchans and crescentic dunes to parabolic dunes. Warren (1970) demonstrated in the Sudan that sand sheets were coarser and less well sorted than crescentic dunes, which were in turn coarser and less well sorted than adjacent linear dunes.

In the Namib Sand Sea (Figure 4.19), there is an improvement in sorting, and an overall decrease in mean grain size from zibars and sand sheets, through barchans and crescentic dunes to linear and star dunes, which tend to be best sorted, although not always the finest. These patterns probably result from the nature of sediment transport on different dunes types. On crescentic dunes, sand movement is essentially unidirectional, and saltating sands are buried on avalanche faces to be recycled later as these deposits are exposed by advance of the dune. Crestal areas of linear dunes are reversed seasonally and undergo more frequent resorting. Star dune crests are best sorted, because they undergo constant reworking by multidirectional winds (Lancaster 1989b).

However the patterns observed in the Namib Sand Sea are not well represented in some other sand seas and dune fields where episodic input of sand from different sources has occurred. In the Gran Desierto, each dune type is associated with sands that are texturally and compositionally distinct (Blount and Lancaster 1990). Similar patterns of grain size and sorting can be observed at Kelso Dunes, where there are considerable differences between

Figure 4.19 Bi-variate plots of grain size and sorting parameters showing comparisons between dune types in the Namib Sand Sea.

DUNE SEDIMENTS

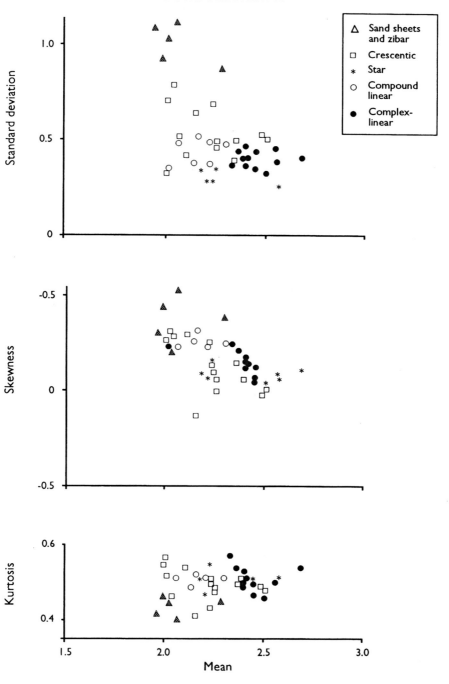

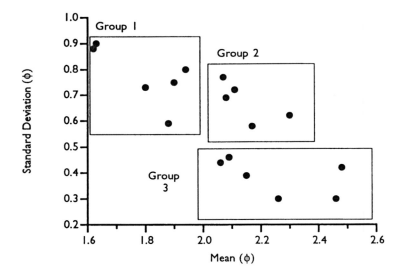

Figure 4.20 Bi-variate plots of grain size and sorting parameters showing comparisons between different dune generations at Kelso dunes.

dunes of each distinct morphologic unit, which represent multiple episodes of sediment input to the dune field (Lancaster 1993b) (Figure 4.20). In addition, sands from active dune areas are consistently finer and better sorted than those from vegetation stabilized dunes (Paisley *et al.* 1991) (Figure 4.21).

General models for grain size and sorting patterns on dunes

Two main grain size and sorting patterns characterize desert dunes: (1) dune crests are consistently finer, better sorted and less positively skewed than lower stoss or plinth and interdune areas; and (2) dune crests are coarser, but better sorted than dune flanks and interdunes. Most aeolian sands can be viewed as a mixture of varying proportions of saltation, reptation and creep, and suspension populations (Visher 1969; Folk 1970). Saltating sand is commonly in the size range 125–250 µm, the modal size group in most aeolian sands. Reptating grains are commonly coarser than 350 µm, and the suspension population is finer than 90 µm. The latter group is poorly represented or absent away from source areas and in all active dune areas. It is, however, quite significant in many vegetation stabilized or relict dunes such as those in the Gran Desierto, Mexico, and at Kelso (Lancaster 1993b) (Figure 4.21).

116

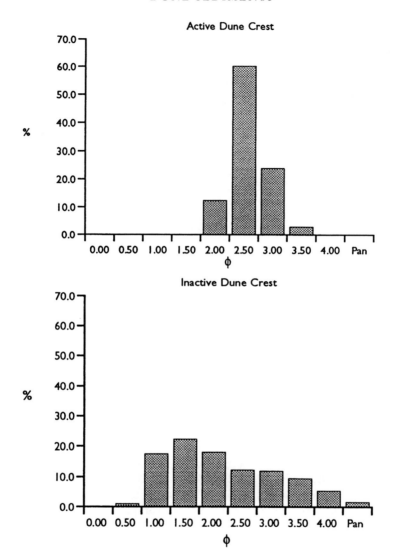

Figure 4.21 Comparison of particle size distributions between active and inactive dunes at Kelso dunes, Mojave Desert.

The grain size and sorting patterns that will evolve will depend on the nature of the source sediments and on the way in which sand transport responds to dune slopes. As sand comes on to the windward slope of dunes, the coarser, reptation load slows down as the slope increases and saltation efficiency decreases. Towards the crest, the slowly moving creep and reptation load is steadily left behind, resulting in a progressive fining of sands towards the crest. On the slip face, avalanching leads to the preferential downslope movement of coarser grains which accumulate at its base. In multidirectional wind regimes, the direction of these processes is reversed, leading to the accumulation of coarser grains on the plinths of many linear and star dunes.

INTERDUNE DEPOSITS

The extent and significance of interdune deposits in modern and ancient aeolian sedimentary environments is being increasingly recognized (Ahlbrandt and Fryberger 1981; Fryberger *et al.* 1983; Kocurek 1981; Hummel and Kocurek 1984; Clemmensen 1989). The area covered by interdune deposits in modern sand seas can be very significant, averaging 50 per cent in the Namib Sand Sea (Lancaster and Teller 1988). The extent and shape of interdune areas varies from one dune type to another. In the Namib Sand Sea the percentage area covered by interdune areas varies from less than 10 per cent in areas of crescentic dunes to as much as 38–73 per cent (mean 60 per cent) for linear dunes. Interdune areas between linear dunes are strongly elongate and extend in a north–south direction for tens of kilometres. They are up to 2 km wide, with a maximum local relief of 10 to 15 m. In areas of star dunes, the interdune areas are more irregular and are typically less than 10 km in length and 1.5 km in width. In areas of crescentic dunes, the slip face of one dune frequently abuts the base of the stoss slope of the next dune downwind, so that interdune areas are restricted to small sub-circular depressions in front of concave slip faces.

Ahlbrandt and Fryberger (1981) recognized two basic types of interdune areas: deflationary (or non-depositional) and depositional. Deflationary interdunes may consist of non-aeolian deposits such as fluvial or alluvial fan sediments as in the northern Namib Sand Sea or the Algodones Dunes (Sharp 1979; Lancaster and Teller 1988), bedrock in the Namib and Akchar Erg of Mauritania (Kocurek *et al.* 1991) or older aeolian sediments, as at White Sands (Ahlbrandt and Fryberger 1981; Simpson and Loope 1985).

Depositional interdunes are divided into wet, damp or dry types. Dry interdune sediments are typically dominated by wind ripple deposits (Plate 19) with localized grainfall and grainflow deposits related to shadow dunes. Many dry interdune deposits are identical to those of sand sheets and are produced by very similar sets of processes. Sedimentary structures of sands in interdune areas between linear dunes in the Namib Sand Sea show that the deposits are composed of flat lying to very gently dipping (2–4°) wind ripple laminations (McKee 1982; Lancaster 1989b). In this respect they differ considerably from interdune deposits in some areas of crescentic dunes, which appear to consist of the bevelled cross-strata of previous generations of such dunes (McKee and Moiola 1975).

Plate 19 Sediments of dry interdune area in the Namib Sand Sea dominated by wind ripple laminae.

Wet interdune deposits form when the interdune areas are flooded by surface water, or where the water table intersects the surface. Wet interdune deposits may include organic deposits (Ahlbrandt and Fryberger 1981), lacustrine carbonates (Lancaster and Teller 1988) (Plate 20), sand deposited by adhesion ripples (Nagtegaal 1971; Hummel and Kocurek 1984), and a variety of evaporite deposits (Ahlbrandt and Fryberger 1981; Fryberger *et al.* 1983).

CONCLUSIONS

Dune sediments, mostly composed of quartz sand derived from fluvial sources, provide a record of dune accumulation. Their characteristics reflect the processes of dune formation and development. A limited number of investigations of dune sedimentary structures has yielded a wealth of information on the fundamental processes of dune accumulation and shown that dunes are composed of varying proportions of wind ripple and avalanche deposits. Whereas most crescentic and many parabolic dunes are composed mostly of avalanche face deposits, those of linear and star dunes are composed of a high proportion of wind ripple deposits. These differences reflect the variations in the primary modes of sediment transport on dunes

119

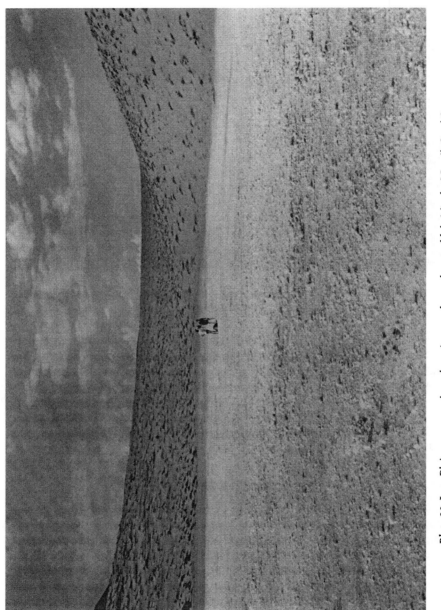

Plate 20 Late Pleistocene wet interdune (seasonal or ephemeral lake) in the Namib Sand Sea.

of different types, as will be expolored in Chapter 5. In addition, in many sand seas, interdune deposits are an important component of the total sediment accumulation.

Spatial variations in sediment transport on dunes give rise to characteristic patterns of sand grain size and sorting so that sand from dune crests tends to be finer and better sorted compared to adjacent interdune and plinth areas. Specific relations between dune types and sediment size and sorting characteristics are however difficult to establish. In many sand seas, sand sheets are often composed of coarse poorly sorted sand, whereas constant reworking of crest areas results in star dune sediments that are much better sorted than other dune types.

5

DUNE DYNAMICS

INTRODUCTION

The initiation, development, and equilibrium morphology of all dunes are determined by changes in sediment transport rates in time and space that give rise to erosion or deposition. Patterns of local erosion and deposition on dunes can be viewed in terms of concepts of sediment continuity. The same principles can also be applied to whole dunes, and even to sand seas, to examine their dynamics in terms of sediment budgets (Figure 5.1). The kinematics of sediment transport require that the mass or volume of sediment is conserved (Middleton and Southard 1984). For each area of the dune surface, a decrease in local sand transport rates with distance means that the influx of sediment to the area will exceed the outflux, leading to sediment storage and therefore an increase in the local bed elevation by deposition of sediment. Conversely, increased sediment transport rates with distance result in sediment outflux exceeding influx, leading to removal of material and lowering of the local bed elevation by erosion of sediment. These relations can be expressed in terms of the sediment continuity equation:

$$dh/dt = -dq_s/dx \qquad (5.1)$$

where h is the local bed elevation, q_s is the local volumetric sediment transport rate in the direction x and t is time.

Spatial changes in sand transport rates are therefore a fundamental control of dune morphology. In areas where wind directions vary seasonally, the spatial pattern of erosion and deposition also changes with each wind season as the dune adjusts towards a new equilibrium with the winds of the time. In turn, the developing bedforms influence local transport rates through form–flow interactions and secondary flow circulations, leading to a dynamic equilibrium between dune morphology and local airflow.

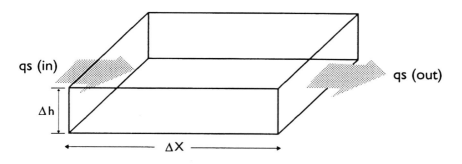

Figure 5.1 The concept of sediment continuity.

DUNE INITIATION

The conditions that lead to dune initiation and the processes involved are poorly understood and little studied, but they clearly are of major importance to understanding how dunes and dune patterns develop. Dune initiation involves localized deposition leading to bedform nucleation, which will then fix a pattern that can propagate downwind (Wilson 1972). Deposition implies a reduction in the local sediment transport rate. This can occur through convergence of streamlines in the lee of obstacles of various sizes; by changes in surface roughness (e.g. vegetation, increased surface particle size) that decrease local shear stress; or by variations in micro-topography (slope changes, relict bedforms).

Airflow around obstacles, such as isolated bushes or shrubs, is characterized by the development of horseshoe vortices that extend downwind on each side of the obstacle (Greeley and Iversen 1985). Beneath the vortices is a zone of increased flow velocity and erosion, but immediately in the lee of the obstacle there is a zone of lower flow velocity and converging, inward directed, vectors of sand transport which lead to deposition (Figure 5.2). Vortices of this type have been observed around roughness elements in the wind tunnel (e.g. Iversen *et al.* 1990) as well as around vegetation (Lee 1991) and boulders. They can occur at different scales and lead to the development of shadow dunes (Hesp 1981; Gunatilaka and Mwango 1989) as well as larger features such as the wind streaks that occur downwind of craters on Mars and Venus and cinder cones on Earth (Greeley and Iversen 1985). Tsoar (1989) has suggested that vegetated linear dunes may develop from a nucleus created by a shadow dune. Development of vortices in the lee of lunette dunes have also been suggested as a mechanism for linear dune formation in the Kalahari and Australia (Twidale 1972; Lancaster 1988b). Larger topographic features such as escarpments and isolated hills may also lead to the development of

A

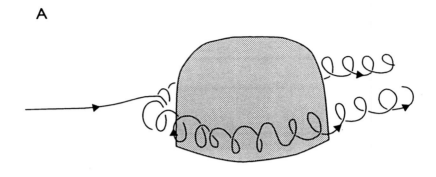

B

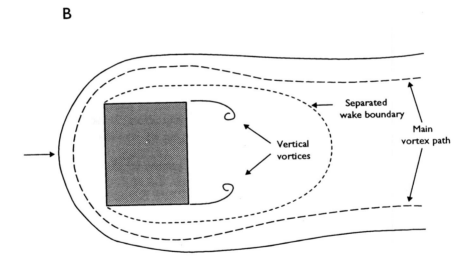

Figure 5.2 Characteristics of a horseshoe vortex around an obstacle (after Greeley and Iversen 1985).

linear dunes in their lee, as at Superstition Mountain in southern California (Smith 1978), and in the Negev (Tsoar 1989). Such dunes often break down into individual barchans where the effect of the topography is no longer important. Falling dunes are a special case of lee dunes in which deposition is concentrated close to the obstacle.

Wind tunnel experiments (Bagnold 1941) suggest that surfaces of mixed grain size (sand and 4 mm pebbles) act as a reservoir in which sand is stored in periods of winds just above threshold, but is removed by stronger winds. Bagnold observed that a sand patch thins and extends downwind in gentle winds (1.1 times u_{*t}), but thickens and extends upwind in strong winds (4 times u_{*t}) provided that a sufficient supply of sand is available. He explained his observations in terms of differences in aerodynamic roughness and threshold shear velocity (u_{*t}) between the surfaces. In periods of gentle winds shear velocity is below threshold for pebbles but not for sand surfaces sand because aerodynamic roughness for the pebble surface is greater than that for a sand surface. Transport is therefore possible only on the sand surface. On the lee side of the sand patch, the increase in aerodynamic roughness as the surface changes from sand to pebbles gives rise to deposition, and the sand patch extends downwind. In periods of strong winds the wind shear velocity exceeds threshold for both pebble and sand surfaces, so that material is eroded from the rough surface. However, the increased quantity of sand in saltation extracts momentum from the wind such that deposition occurs on the sand patch and at its upwind margin, extending it upwind and increasing its thickness.

Recent work by Greeley and Iversen (1987) has shown that for winds of a given speed at height (z), values of u_* are greater on aerodynamically rougher surfaces, but threshold shear velocities also increase on the rougher surfaces by an amount that is related to the ratio between particle diameter and aerodynamic roughness:

$$u_{*tr}/u_{*t} = 2(D_p/z_o)^{-1/5} \tag{5.2}$$

where u_{*tr} is the threshold shear velocity for the rougher surface, and D_p is the diameter of the particles in transport. Transport rates may therefore be larger or smaller on the rougher surface, depending on the relative values of u_*, u_{*t} (smooth) and u_{*t} (rough) and the ratio between particle diameter and z_o. Blumberg and Greeley (1993) show that potential sand transport decreases sharply for increases in aerodynamic roughness between 0.01 and 0.1 cm for particles with a diameter greater than 100 µm (see Figure 2.17). This corresponds to a change from playas and sand surfaces to gravelly alluvial fans.

Greeley and Iversen suggested that transport takes place only on smoother surfaces at low wind speeds; at intermediate speeds transport occurs on both surfaces, with greater transport on the smooth surface. However, at high wind speeds, transport rates are higher on the rough surface, implying that sand patches will erode at low and intermediate wind speeds, but accrete at high wind speeds. This is an alternative explanation for the phenomena observed by Bagnold, but Greeley and Iversen do not take into account the increase in aerodynamic roughness that occurs when saltation takes place (saltation roughness).

When winds pass from surfaces with one roughness to another, the wind profile changes and requires some distance to reach equilibrium with the new surface (Hunt and Simpson 1982). An internal boundary layer develops downwind from the edge of the change in roughness and gradually diffuses up into pre-existing boundary layer. The leading edge of a sand patch will therefore be a zone of fluctuating flow and zones of alternate erosion and deposition. Bagnold (1941) observed in the wind tunnel that this zone

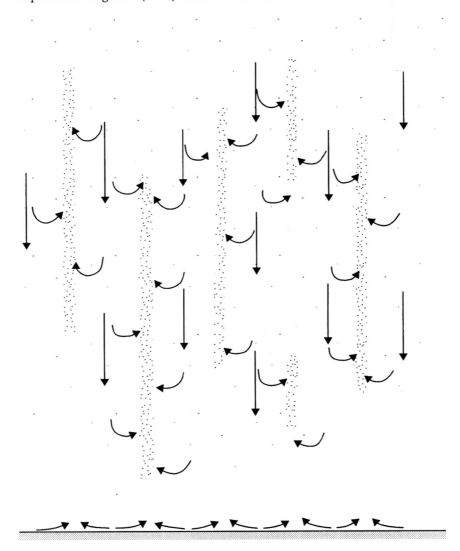

Figure 5.3 Development of sand strips by transverse instability of a sand-transporting wind (after Bagnold 1941).

extended for a distance 4 to 6 m, a length scale that he suggested was the minimum size for a true dune.

Field studies of the initiation and early development of dunes are rare. Cooper (1958) and Jäkel (1980) described the development of barchans and transverse ridges from thin sand patches with no flow separation in their lee to small dunes with lee side flow separation, but did not document how they were initiated. Bagnold (1941) and Madigan (1946) noted the transverse instability of a wind moving over a gravel surface which produced parallel longitudinal wavy ribbons of sand transport. They suggested that these could be deposited as longitudinal sand strips when the wind speed dropped and, under favourable circumstances, might ultimately develop into linear dunes, as surface roughness differences would encourage deposition on the sandy area (Figure 5.3).

Kocurek *et al.* (1992) have documented the initiation and growth of dunes at Padre Island, Texas. Nucleation of bedforms takes place where wind speeds are reduced by 37 to 86 per cent of upwind values due to changes in aerodynamic roughness or micro topography as a result of patches of grass, erosional depressions, relict dune topography, or vegetated sand ridges (shadow dunes). Not all sites of initial deposition develop into dunes, but growth is favoured by a greater initial amount of deposition and enhanced local sand supply. Many small protodunes lose sand in high wind events and some become deflated away entirely. Kocurek *et al.* (1992) recognized five stages of dune initiation and development (Figure 5.4) with a progressive evolution of the lee face and bedform induced secondary flow expansion and separation: (1) irregular patches of dry sand a few cm high, (2) 0.1–0.35 m high protodunes with wind ripples on all surfaces, (3) 0.25–0.40 cm proto-dunes with grainfall on lee slopes, (4) 1–1.5 m high barchans with grainflow, and (5) 1–2 m high crescentic ridges. The developing dunes were characterized by a reverse asymmetry (steeper stoss slope) in stage 2, similar to that noted by Cooper (1958). The change from flow expansion to flow separation came as lee slopes exceeded 22°. However, not all dunes develop at the same rate, and retrograde evolution occurs in some cases. Once dunes have reached stage 4 or 5, subsequent evolution of the dune field mainly takes place by repeated merging, splitting, lateral linking and cannibalization of dunes. Dune growth is at the expense of interdune areas and the pattern evolves so that the height/spacing ratio tends toward 1/20.

AIRFLOW OVER DUNES

As dunes grow, they project into the atmospheric boundary layer. Airflow is therefore compressed so that streamlines converge towards the crest, giving rise to a corresponding increase in wind speed and shear stress on the stoss, or windward slope (Figure 5.5). These phenomena are of fundamental importance in understanding the morphology of all dunes. On the lee side,

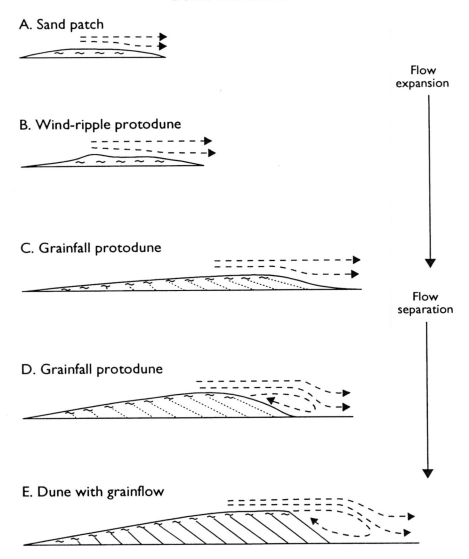

Figure 5.4 Stages of dune initiation (after Kocurek *et al.* 1992).

flow separation, expansion, and divergence lead to a variety of secondary flow patterns. In multidirectional wind regimes, the nature of interactions between dune form and airflow changes as winds vary in direction seasonally, and major lee side secondary flow patterns become important in determining dune morphology and dynamics.

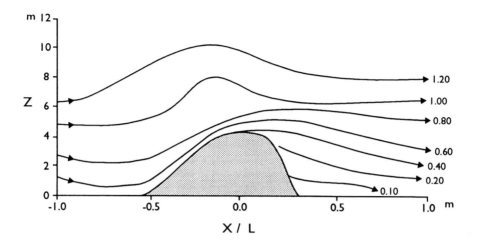

Figure 5.5 Pattern of streamline convergence over a barchan dune (after Lancaster 1994a).

Theoretical considerations

Airflow over dunes can be represented in a similar way to flow over low hills, using analytical models of turbulent flow. These models follow the work of Jackson and Hunt (1975) and Hunt *et al.* (1988) and divide the boundary layer into inner and outer regions (Figure 5.6). The inner region has a thickness l (about 1/20 the characteristic length L) in which turbulence is approximately in equilibrium and shear stress changes significantly affect the mean flow, allowing use of mixing length closure models. The inner layer consists of two sub-layers: a very thin inner surface layer (ISL) with a thickness approximately equal to the size of the roughness elements, and a shear stress layer (SSL) in which the effects of shear stress changes decrease away from the surface. Above the inner region, changes in shear stress have small effects on the mean flow over the obstacle. The outer layer flow can be treated as inviscid and irrotational, whereas flow in the shear stress layer is affected by the upwind shear, especially when the atmosphere is stable. Following Jackson and Hunt (1975), the inner layer thickness (l) is given by:

$$l/z_o = 1/8 \, (L \, z_o)^{0.9} \tag{5.3}$$

where L is the half length of the hill (see Figure 5.5). Frank and Kocurek (1994b) and Lancaster *et al.* (1994) estimate l to be 0.5–1.0 m for a range of small crescentic dunes.

The basic predictions of this type of model have been confirmed by field experiments (for a review see Taylor *et al.* (1987)) and numerical simulations

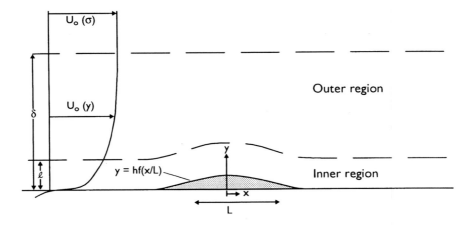

Figure 5.6 Model for wind-flow over a low hill (after Jackson and Hunt 1975).

(Hunt and Simpson 1982), and provide a useful prediction of some characteristics of airflow over dunes, especially velocity amplification on the stoss slope (see below). Versions of this model have also been used to simulate airflow over dunes (e.g. Walmsley and Howard 1985; Weng *et al.* 1991).

The stoss slope

Increases in wind velocity on the stoss slopes of dunes are predicted by the models of wind flow over hills and escarpments (Jackson and Hunt 1975; Bowen and Lindley 1977; Pearse *et al.* 1981; Norstrud 1982). Measurements of wind speed over crescentic, linear and star dunes by many workers show that wind speeds at dune crests are typically 1.1 to 2.0 times those measured immediately upwind of the dune (Howard *et al.* 1978; Lancaster 1985b, 1987; Tsoar 1985; Livingstone 1986; Mulligan 1988; Burkinshaw *et al.* 1993; Wiggs 1993; Lancaster *et al.* 1994). Models (e.g. Tsoar 1985) and field data also indicate a drop in wind velocity immediately upwind of the dune. The rate of velocity increase on the stoss slope is not linear (Figure 5.7), and follows the slope of the dune closely, decreasing towards the crest (Tsoar 1985; Mulligan 1988). Velocity increases of a similar magnitude have been measured over small hills (see Taylor *et al.* (1987) for a recent review). The magnitude of the velocity increase is represented by the speed-up ratio (Δs) or amplification factor (Az). Thus:

$$Az = U_2/U_1 \qquad (5.4)$$

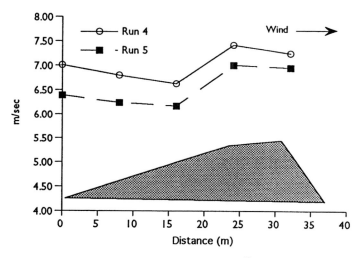

Figure 5.7 Increase in wind velocity on the stoss slope of a barchan dune (after Lancaster *et al.* 1994).

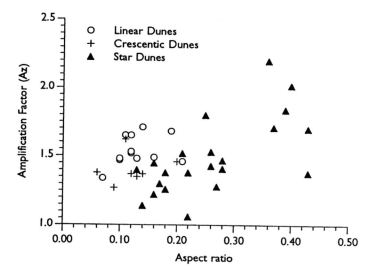

Figure 5.8 Relation between amplification factor and dune aspect ratio. Data from Namib linear dunes and Gran Desierto star dunes (after Lancaster 1994a).

where U_2 is the velocity at the dune crest and U_1 is the velocity at the upwind base of the dune. The amplification factor varies with dune height and its aspect ratio (the steepness of the dune, h/L) (Lancaster 1985b; Tsoar 1985) (Figure 5.8).

131

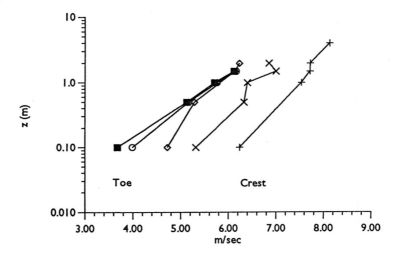

Figure 5.9 Wind velocity profiles on the stoss slope of a barchan dune (after Lancaster *et al.* 1994).

Measurements of winds on linear and star dunes by Tsoar and Lancaster show that amplification factors also vary directly with the angle of attack of the wind, or the wind direction relative to the dune. This is because the effective aspect ratio increases as the wind blows more nearly perpendicular to the dune crest (Tsoar 1989). Thus the effective aspect ratio becomes h/La where

$$La = L/\sin \alpha. \tag{5.5}$$

where La is the dune length measured parallel to the wind direction, L is the dune length, and α is the angle of attack.

As Bagnold (1941) and Watson (1987) have pointed out, the equilibrium morphology of dunes is controlled by the pattern of wind shear velocity (u_*) over the dune because it is u_* that determines sediment transport rates. Wind velocity profiles have been measured on the stoss slopes of dunes by many workers (Howard *et al.* 1978; Mulligan 1988; Weng *et al.* 1991; Burkinshaw *et al.* 1993; Wiggs 1993; Frank and Kocurek 1994b; Lancaster *et al.* 1994). These studies show that profiles are approximately log-linear near the base of the slope, but become progressively less so up slope (Figure 5.9), as a result of the effects of flow acceleration and streamline convergence. Unlike Wiggs (1993), Frank and Kocurek (1994b) and Lancaster *et al.* (1994) recognize a decrease in wind shear stress up the dune slope, even though wind speed and measured sediment transport increase up the dune (Figure 5.10). They

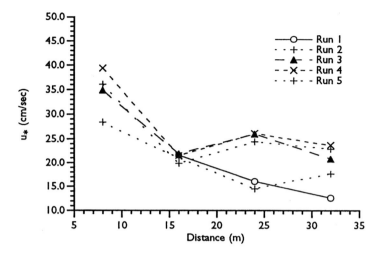

Figure 5.10 Pattern of measured wind shear velocity on the stoss slope of a barchan dune (after Lancaster *et al.* 1994).

suggest that this phenomenon results from the progressive development of an internal boundary layer equivalent to the inner layer of Jackson and Hunt (1975) on the dune slope (Figure 5.11). The inner layer, with a maximum thickness of 0.5–1.0 m, is characterized by increasing shear velocity, but has not been recognized because conventional wind profiles derived from anemometry on dunes do not measure the part of the boundary layer that is significant for sediment transport.

Recognizing the considerable problems associated with derivation of wind shear velocity and sediment transport rates from wind profiles on dune slopes, Lancaster *et al.* (1994) measured sediment flux rates directly using an array of sand traps on the stoss slope of a 5 m high barchan dune. Their data indicate an increase in flux up the dune, with the maximum total and mean flux occurring at the brink line (Figure 5.12). Similar observations were made by Wiggs (1993). The shape of the flux vs distance curve also varies with wind speed conditions. At low wind speeds, sediment flux tends to increase exponentially with distance up the dune. At higher wind speeds, the increase becomes approximately linear.

This pattern probably reflects both a threshold entrainment equilibrium condition and a fetch effect. When wind velocities are low and close to threshold at the dune toe, saltation flux initially increases slowly with distance until wind velocity accelerates sufficiently to maintain a fully saturated saltation flux. This condition is only reached well up the stoss slope. By contrast, when wind velocities are high, saturated flux is reached

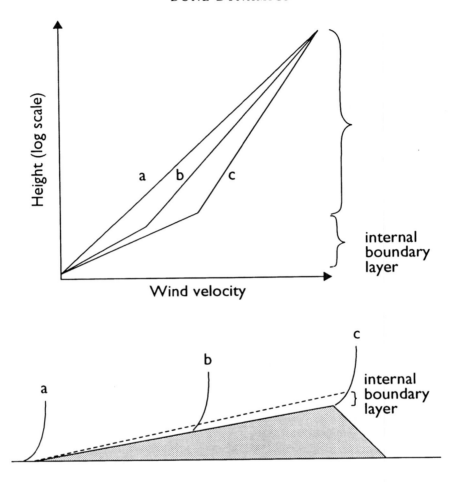

Figure 5.11 Development of internal boundary layers on the stoss slope of a flow-transverse dune (after Frank and Kocurek 1994b).

in a short distance and then increases linearly with distance as the airflow accelerates up the dune. However, it is unlikely that equilibrium between saltation flux and the wind profile is ever established because of the continuous flow acceleration up the dune.

The ratio between mean flux at the brink and toe of the dune decreases linearly with wind shear velocity measured at the dune toe (Figure 5.13). These observations support the model of Lancaster (1985b) that suggests that differences between sand transport rates at the base and crest of the stoss slope will be at a maximum when incident winds are just above threshold

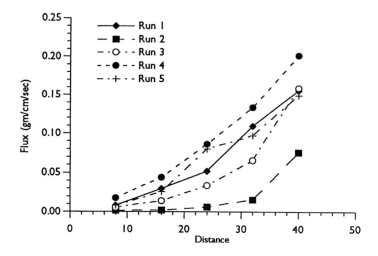

Figure 5.12 Increase in measured sand flux on the stoss slope of a barchan dune (after Lancaster *et al.* 1994).

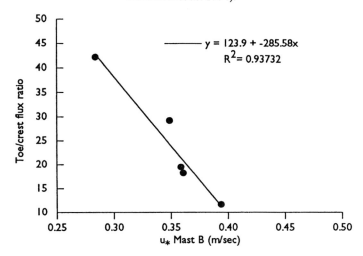

Figure 5.13 Relations between the crest/toe ratio of measured sediment flux and incident wind shear velocity (after Lancaster *et al.* 1994).

velocity, and decrease as the overall wind speed increases. This implies that sand transport will occur only in the crestal areas of dunes in periods of low wind speeds, whereas in periods of strong winds all of the dune will be mobile. These patterns can be observed on many dunes, and are especially important on large dunes. The crest may be eroded and the dune lowered in

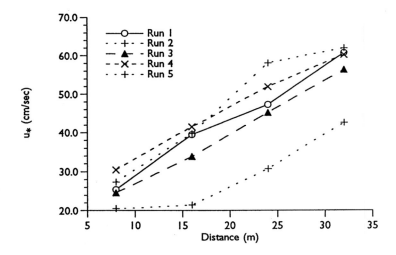

Figure 5.14 Wind shear velocity pattern derived from flux measurements (Figure 5.12) using a simple Bagnold sand-transport equation (after Lancaster *et al.* 1994).

periods of low wind speeds when sand transport is confined to the crest, but grow in periods of stronger winds which can transport sand to crestal areas.

Data on sediment flux can also be used to estimate the magnitude of the wind shear velocity required to generate the observed sediment flux using a simple Bagnold flux equation with the measured flux as an input variable. The pattern of u_* values generated in this way is shown in Figure 5.14 and indicates that u_* should increase up the stoss slope of this dune by 2–3 times from toe to crest. This increase in u_* is comparable to that predicted by Weng *et al.* (1991) from numerical model studies, and confirms model predictions and experimental data (e.g. Lai and Wu 1978; Walmsley and Howard 1985; McLean and Smith 1986) that indicate a shear stress maximum on the upper stoss slope of flow-transverse bedforms (Figure 5.15).

The lee slope

In the lee of the crest, wind velocities and transport rates decrease rapidly as a result of flow expansion between the crest and brink and flow separation on the avalanche face. Sweet and Kocurek (1990) suggest that there are three types of flow in the lee of dunes: (1) separated, (2) attached, and (3) attached and deflected (Plate 21). The nature of lee side flow is controlled by the dune shape (aspect ratio), the incidence angle between the primary wind and the crestline (the angle of attack), and the stability of the atmosphere (Figure 5.16). The dune shape and angle of attack vary along the dune crestline, giving rise to important spatial variations in the nature of lee side airflow,

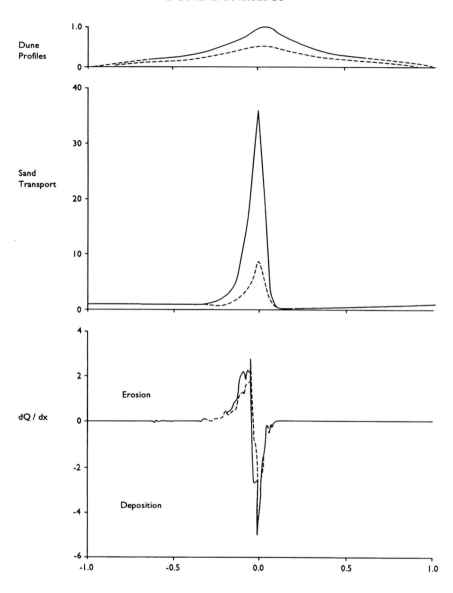

Figure 5.15 Simulated pattern of wind shear velocity and sediment flux
(after Lai and Wu 1978).

(a)

(b)

Plate 21 Dune lee side processes: (a) Avalanching. (b) Lee side flow separation. (c) Along dune transport reworking grainflow deposits.

even with the same incident wind speed and direction. This is especially important on linear and star dunes.

Flow separation occurs mainly where the dune aspect ratio is high and flow is transverse to the crest. Wind speeds at the base of the slip face of crescentic dunes are low and fluctuating. They average 0.40–0.80 of those on adjacent crests in some Namib examples, and as little as 0.04 of crest values for crescentic dunes at Padre Island, Texas (Sweet and Kocurek 1990). The existence of eddies and reversed flow directions is controversial, due in part to the considerable problems of visualizing such flows. Reverse flow in the lee of crescentic dunes was discounted by Cooper (1958) and Sharp (1966), but field observations by Hoyt (1965), Warren and Knott (1983), Sweet and Kocurek (1990) and Frank (forthcoming) support its existence. The width of the separation zone for dunes that are truly transverse to the flow is controlled by the wind speed at the crest, dune aspect ratio, and atmospheric stability, but never exceeds two dune heights according to Sweet and Kocurek (1990). However, Lancaster (1989b) notes that wind velocities in the lee of isolated crescentic dunes do not recover to upwind values before a distance equivalent to 10–15 dune heights is reached. Detailed studies of lee side airflow by Frank (in press) show that the reverse flow eddy extends about four dune heights downwind of the dune brink, with a range in the point of reattachment from 1.6 to 5.4 dune heights. The size of the separation zone is comparable to that measured over sub-aqueous bedforms (Engel

A

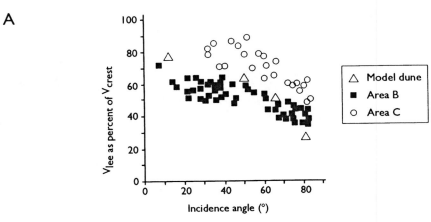

B

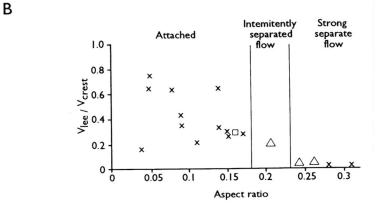

C

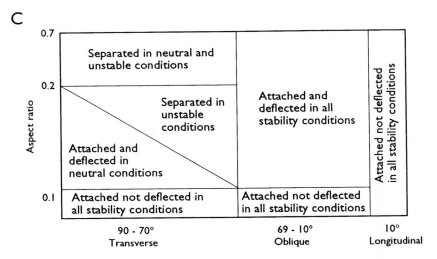

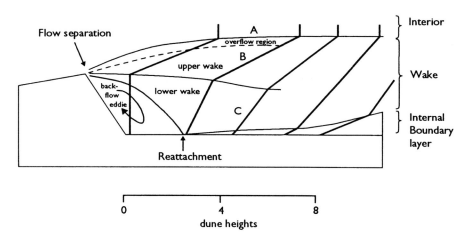

Figure 5.17 Model for airflow on the lee side of flow-transverse dunes
(after Frank in press).

1981; Nelson and Smith 1989). Beyond the point of reattachment, the wind
velocity profiles can be divided into four regions (Figure 5.17): (1) the
interior is a low shear region above the dune where the effects of the dune
are minimal; (2) a transition zone with high shear, because the upper part of
this zone exhibits decelerating flow while the lower part consists of the upper
wake with accelerating flow; (3) the lower wake with low shear and low wind
speeds; and (4) the internal boundary layer, which is characterized by
relatively high shear. Downwind from the dune, the wake is progressively
accelerated as momentum is transmitted downward from the airflow above
the dune. This results in increased shear in the lower part of the wake so that
by about eight dune heights from the brink, the lower wake and transition
zone have merged into a single zone in which the profile is log-linear.

Winds that approach the dune crest at an angle are diverted to flow parallel
or sub-parallel to the lee face. The degree of wind deflection is in all cases a
cosine function of the incidence angle between the crestline and the primary
wind. Thus:

$$U_l = U_c \left(K \, ln \left(\cos \alpha + A \right) \right) \tag{5.6}$$

Figure 5.16 Flow on the lee side of dunes (modified from Sweet and Kocurek 1990).
A: reduction in wind velocity in lee slope as a function of incidence angle; B:
reduction in wind velocity in lee slope as a function of dune aspect ratio; C:
generalized model of lee slope flow.

where U_l is the wind speed at the base of the lee slope, U_c is the wind speed at the dune crest, α is the incidence angle, A is the maximum value of U_l/U_c and K is an empirical constant that is 0.83 for Algodones Dunes (Sweet and Kocurek 1990). The degree of deflection is therefore inversely proportional to the incidence angle between the crestline and the primary wind. The magnitude of the deflected flow can be described by:

$$Cp\,(z) = V_w\,(z + dz)\cos\alpha \qquad (5.7)$$

where $Cp\,(z)$ is the magnitude of the wind vector parallel to the crest, V_w is the magnitude of the incident wind at the crest, α is the incidence angle for the wind, z is the measurement height and dz is the change in elevation as a result of flow separation. Relatively high lee side wind velocities are associated with low aspect ratio dunes and oblique primary winds. When the angle between the dune and the wind is less than 40° the velocity of the deflected wind is greater than that at the crest and sand is transported along the dune (Tsoar 1983a; Nielson and Kocurek 1987; Lancaster 1989a). When winds are at an angle of more than 40° to the crestline the velocity of the deflected wind will be reduced, giving rise to lee side deposition.

Lee slope airflow on linear and star dunes

Airflow in the lee of linear and star dunes is complex, especially where the crestline is sinuous. The same primary flow may be simultaneously separated or deflected at different points along the dune. The detailed measurements of winds and sand movements on a 6 to 13 m high sinuous simple linear dune by Tsoar (1978, 1983a) show that a separation zone exists in the lee of the

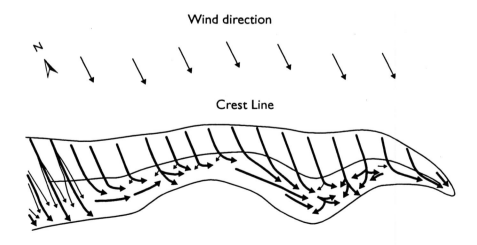

Figure 5.18 Patterns of lee side airflow on a linear dune (after Tsoar 1978).

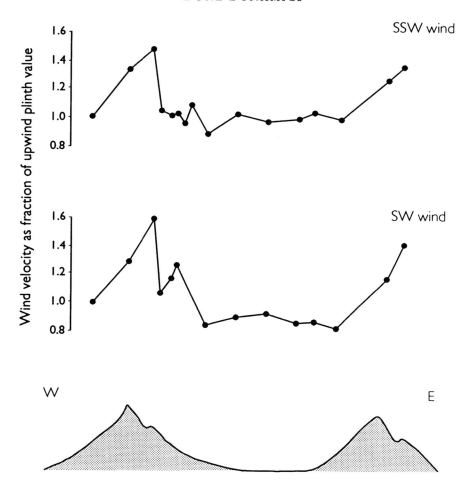

Figure 5.19 Pattern of wind velocity over a Namib linear dune and interdune (after Lancaster 1989b).

crest even when winds blow obliquely to the dune, which has a high aspect ratio resulting from its triangular profile (Figure 5.18). In this zone, wind velocities are 0.20–0.70 of those at the crest. Surface winds in this zone are erratic, but often are directed towards the crest, indicating that there is a small lee helical eddy in the separation zone. At the point of attachment of the separated flow, 5–7 m from the crest, wind velocities increase again, and may reach values 1.05–1.20 of those measured at the crest. Wind directions observed in this zone are parallel to the dune. The increased velocities and sand transport rates in the attachment zone are the result of the concentration of streamlines as the separated flow returns to the dune surface. The greatest

East flank dunes   Main crest

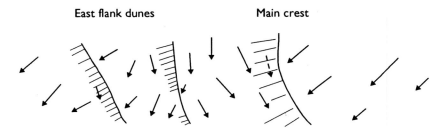

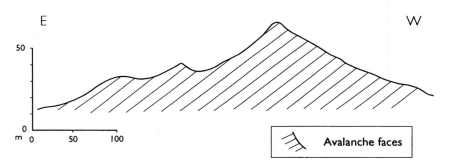

Figure 5.20 Pattern of wind direction over a Namib linear dune (after Lancaster 1989b).

increase in wind velocity parallel to the dune occurred when the wind crossed the crest at an angle of 30° ± 10° to the dune. Beyond this zone of reattachment, wind velocities tend to decrease again and to resume their original direction (Figure 5.18).

Observations by Livingstone (1986) and Lancaster (1989b) show that winds oblique to the trend of complex linear dunes in the Namib separate at the crest and wind velocities at the base of the main slip face are only 0.50–0.70 of those at the adjacent crest. Wind speed then increases slightly in an irregular manner over the east flank crescentic dunes, but drops again on the east plinth, and remains approximately constant over the interdune area

Figure 5.21 Patterns of wind directions on a star dune (after Lancaster 1989a). A: winter northerly winds; B: spring westerly winds; C: summer southerly winds.

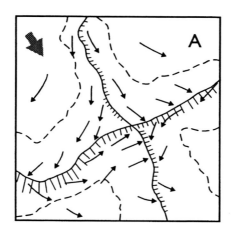

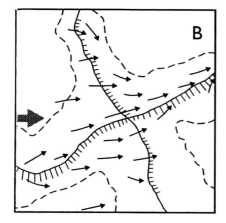

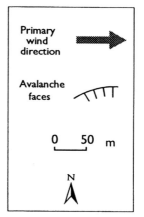

Primary wind direction

Avalanche faces

0   50 m

N

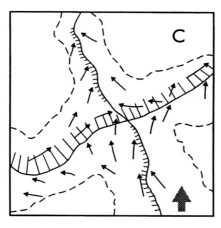

(Figures 5.19 and 5.20). Smoke candles and wind ripple patterns (Figure 5.20) show that the pattern of wind directions observed on large complex linear dunes in the Namib Sand Sea is very similar to that observed by Tsoar, and indicates that similar processes operate on linear dunes of all sizes. However, the zone of lee side airflow diversion extends up to 100 m from the main dune crest, approximately four times the height of the avalanche face, or one tenth the total dune width. This compares with a 5 to 7 m wide flow diversion zone observed by Tsoar on a dune which averaged 11 to 13 m high. In this case, the zone of flow diversion covers most of the lee flank of the dune, whereas in the Namib example, flow diversion affects only the upper parts of the dune. As a result, sand can leave the dunes in periods when wind velocities are high enough to promote sand transport on plinths and be transferred from one dune to another (Livingstone 1986).

Airflow on the star dune studied by Lancaster (1989b) is directed up the windward arm of the dune, but radially outward on the stoss slopes of both flow-transverse arms. There is a strong separation of flow at the crest of both east and west arms of the dune (Figure 5.21), giving rise to a zone of very low wind speeds immediately in the lee (0.10–0.40 of crest values). Beyond this zone a well-developed secondary circulation in the lee of these arms sweeps sand along the base of the avalanche face towards the flow-parallel arm. This circulation is a return flow, probably in the form of a helical eddy, from the end of the dune arm towards the low pressure zone created by flow separation at the dune crest. The pattern is reversed as wind changes direction seasonally.

EROSION AND DEPOSITION PATTERNS ON DUNES

The pattern of erosion and deposition on a dune provides information on how it behaves under different wind conditions, so that the processes that maintain its form can be inferred.

Crescentic dunes

Erosion and deposition patterns on crescentic dunes are characterized by erosion on the stoss slopes and deposition in the lee. This pattern leads to a migration of the dune downwind. The maximum deposition occurs 0.2–0.4 m from the top of the lee face (Anderson 1988), and declines exponentially with distance from this point (Figure 5.22). The point of maximum deposition is the result of all saltation trajectories that leave any point on the dune upwind of the brink with sufficient momentum to travel the intervening distance (Anderson 1988). The decrease in deposition rate has a length scale of approximately 1 m, typically less than the length of the lee face, giving rise to oversteepening and subsequent failure and avalanching.

Almost all sand transported over the crest of crescentic dunes is deposited,

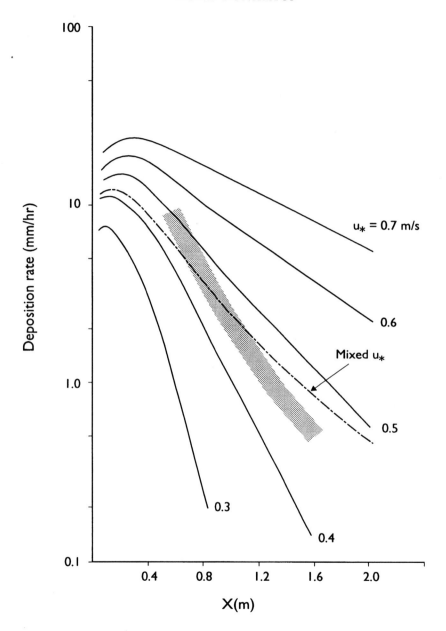

Figure 5.22 Calculated pattern of deposition on a dune lee face for selected wind shear velocities at brinkline and a range of particle sizes around a mean of 300 μm. Shaded area is data from Hunter 1985 (after Anderson 1988).

so that they are typically 'sand trapping' bedforms (Wilson 1972). The movement of crescentic dunes can be described by:

$$c = Q/yh \tag{5.8}$$

where c is the migration rate, Q is the bulk volumetric sand transport rate, y is the bulk density of sand and h is dune height (Bagnold 1941). An inverse relationship between dune height and migration rate (Figure 5.23) has been shown by numerous investigators (e.g. Finkel 1959; Long and Sharp 1964; Hastenrath 1967; Tsoar 1974; Embabi 1982; Endrody-Younga 1982; Slattery 1990).

The amount of erosion and deposition occurring on a dune surface is proportional to the slope angle and the rate of advance of that part of the dune (Bagnold 1941; Allen 1968b). Thus

$$dq/dx = y\, c \tan \alpha \tag{5.9}$$

where dq/dx is the rate of sand removal or deposition per unit length at any point, y is the bulk specific weight of dune sand, c is the rate of advance of the dune and α is the slope angle. There will be no erosion at the dune crest, and when $\tan \alpha$ is negative, as on lee slopes, dq/dx is also negative and thus deposition will occur. This relation is however a purely geometric one, and does not consider the effects of changing wind velocities and sand transport rates over the dune on erosion and deposition rates or the rate of advance of the dunes.

Few direct measurements of erosion and deposition have however been

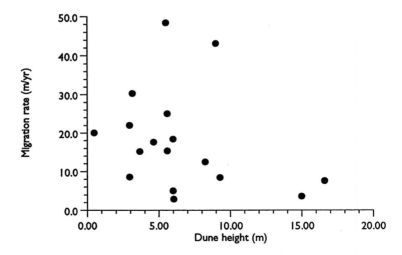

Figure 5.23 Relations between dune height and migration rate. Data for crescentic ridges and barchans (after Lancaster 1994a).

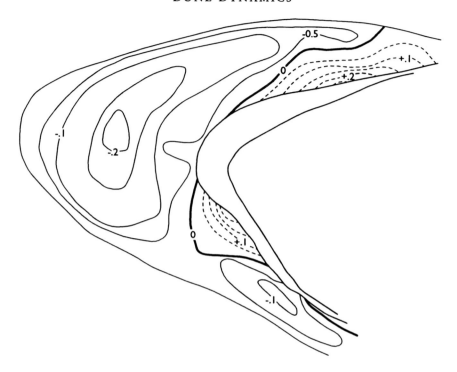

Figure 5.24 Changes in surface elevation on a barchan dune (after Howard
et al. 1978).

made on crescentic dunes. Howard *et al.* (1978) measured changes in surface
elevation on a barchan dune over a two-week period (Figure 5.24) and found
an erosion maximum on the central part of the stoss slope and on the
southern horn, with deposition on brink each side of the crestal area of the
dune. This pattern agrees well with that predicted by a transport rate
simulation model which includes the effects of the interaction of winds with
the dune topography.

Linear dunes

All linear dunes display a similar pattern of erosion and deposition which is
characterized by low activity on the plinth and high rates of erosion and
deposition on the upper flanks of the dune with a peak at the dune crest and
major slip face.

Tsoar (1978, 1983a) measured changes in the cross-section of a simple
linear dune and found that the crestline migrated laterally over a distance of

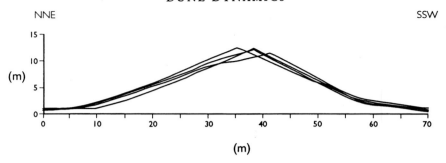

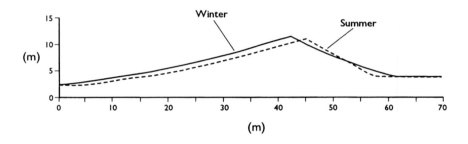

Figure 5.25 Changes in the profile form of a simple linear dune with seasonally varying wind directions (after Tsoar 1985).

5–7 m under the influence of summer NNE and winter SSW–SW winds so that the dune became more asymmetric as the wind season progressed (Figure 5.25). Seasonal changes in the wind directions give rise to spatial changes in the pattern of erosion or deposition on the dune, creating a sinuous pattern of narrower, lower saddles and higher, wider peaks (Figure 5.26). This then maintains itself by interaction between the dune and the wind. For winds from either side of the dune, the crestline will consist of flow-parallel and flow-transverse segments, with erosion on the former and deposition on the latter. The dune will narrow when erosion exceeds deposition, other areas will be sites of net deposition and widen and grow in height. However, as deposition will be spread over a greater area, the growth rate will progressively decrease and eventually a dynamic equilibrium state will exist, with peaks and saddles at a uniform equilibrium height. Tsoar suggested that the overall tendency will be for the dune to extend downwind, with the sinuosities gradually moving along the dune. Measured rates of extension on the dune studied by Tsoar averaged 14.52 m per year. In the Sinai, the overall trend of the dune is determined by the summer wind (northeasterly), which crosses the dune at the optimal angle for lee flank erosion and movement along the dune. Winter winds (southwesterly) in the

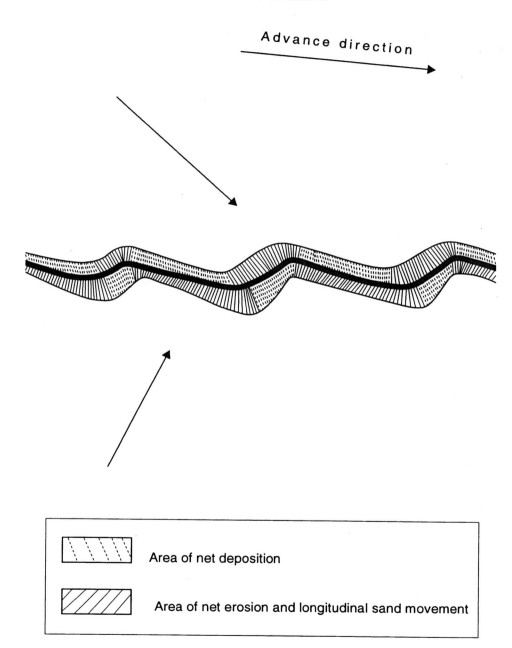

Figure 5.26 Patterns of erosion and deposition along a simple linear dune in changing wind directions (after Tsoar 1983a).

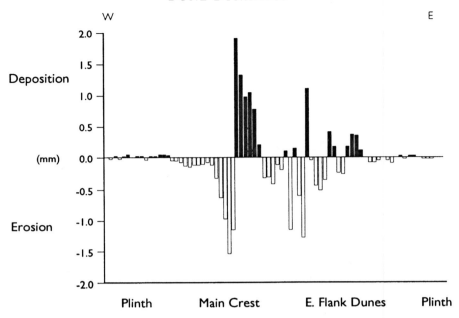

Figure 5.27 Patterns of annual net erosion and deposition on a Namib linear dune (after Lancaster 1989b).

Sinai also move sand along the dune and reverse the position of erosion and deposition areas, but do not cross the dune at optimal angles and so contribute less to its elongation.

Large S–N trending complex linear dunes in the Namib Sand Sea occur in a wind regime that has two distinct seasonal directional sectors: S to SW in the period September–April, and NE to E in May to August. Studies by Livingstone (1986, 1989) and Lancaster (1989b) show that the magnitude of erosion and deposition varies significantly from place to place across linear dunes (Figure 5.27). Fifty to seventy per cent of the total of erosion and deposition takes place in crestal areas with a further 18 to 30 per cent on east flank barchanoid dunes.

Erosion and deposition on dune plinths is at a low level at all seasons, with an irregular spatial and temporal pattern of small-scale erosion and deposition. The crestal regions of the dunes are the most active as they are reworked by winds from different directions according to season (Figure 5.28). The crestlines migrate over a lateral distance of as much as 14 m over a 12-month period but with little net change over a period of 2 to 3 years (Livingstone 1989). However, on a decadal time scale, significant changes in form do occur in the crestal areas of these dunes (Livingstone 1993) (Figure 5.29).

From October to March, crestlines migrate towards the east or northeast

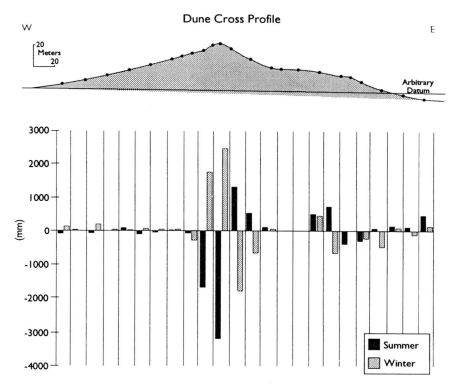

Figure 5.28 Seasonal changes in pattern of erosion and deposition on a Namib linear dune corresponding to changes between summer SSW to SW winds and winter easterly winds (after Livingstone 1989).

in a period of dominantly SSW to SW winds, reaching their maximum easterly position in early March. There is then little change in crest position until the season of NE to E winds, which erode the upper parts of the slip face and deposit sand as a new slip face on the western side of the crest, which migrates simultaneously towards the west or southwest. An increase in dune height during this period was documented by Livingstone (1989). Erosion and deposition patterns on the upper east flanks of the dunes are complex as a result of infilling or scour of surface irregularities. Most east flank barchanoid dunes recorded a net movement towards the northeast, of as much as 5 m.yr^{-1} (mean 2.99 m.yr^{-1}), with the advance rate being inversely related to the height of the dunes.

The overall trend of linear dunes in the Namib Sand Sea appears to be determined by the persistent SSW–SW winds that cross dune crests at optimal angles for lee side flow diversion and transport along the dune. The infrequent, but high energy, easterly winds reverse the dominant pattern of

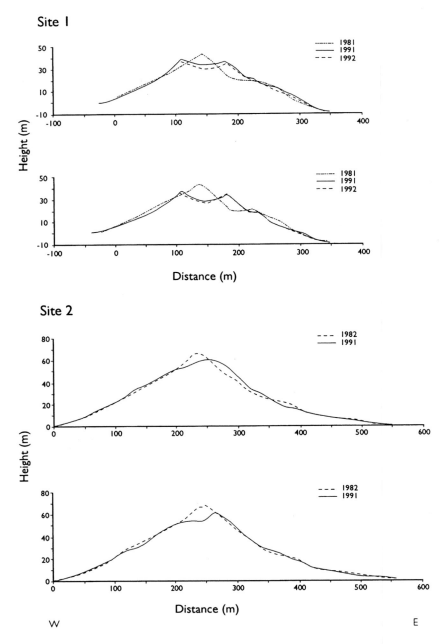

Figure 5.29 Changes in the crestal areas of a linear dune in the Namib over a 10-year period (after Livingstone 1993).

sand transport, but cross dune crests at an angle that tends to promote lee side deposition, so that sand tends to stay on the dune (Livingstone 1988).

Star dunes

Observations of patterns of erosion and deposition on a 40 m high star dune in the Gran Desierto Sand Sea indicate that most of the change in dune form is associated with the reworking of the crestal areas of the main NE–SW oriented arms of the dune as winds change direction seasonally (Lancaster 1989a). Similar observations were made by Sharp (1966) on reversing dunes at Kelso, in the Mojave Desert. In the Gran Desierto, winds are from three major sectors. Winter northerly, spring westerly and summer southerly winds each generate 25–30 per cent of annual potential sand transport. Approximately three times as much erosion and deposition takes place on the east–west arms which are approximately transverse to the major wind directions, compared with the north or south arms of the dune, which are parallel or oblique to these winds (Figure 5.30). The plinths experience relatively little change in all seasons, whereas the crestlines of the east and west arms migrate over a distance of as much as 20 m. In the course of each major wind season, the profile shape of the main crestline changes (Figure 5.31). The initial state corresponds to the asymmetric profile developed in the previous wind season, with a convex stoss slope profile and a lee slope at the angle of repose. During the early part of the new wind season, the upper part of the former lee face is eroded and a small avalanche face develops on the upper part of the new lee face. The crestline is sharp and the dune tends to gain height at this time of year. As the wind season continues, the crestline migrates downwind and the avalanche face prograded over the former stoss slope, gradually increasing in height as it does so. The dune crestline passes through a period when it is near symmetrical in profile, but becomes progressively more asymmetric in the latter part of the season. Erosion rates are highest at the crest of the dune soon after winds change direction, but decline thereafter, as the point of maximum erosion moves upwind and the dune becomes more fully adjusted to the new conditions. The rate of advance of the crestline slows down as the wind season continues and the avalanche face height increases.

The spatial patterns of erosion and deposition change seasonally, as the primary incident wind direction changes (Figure 5.30). During the winter season of northerly winds, erosion is concentrated in the northwestern quadrant of the dune, on the upper slopes of the east arm, and on the western slope of the south arm. Deposition dominates the northeastern and south-eastern quadrants of the dune, and on the south-facing avalanche faces along the east and west arms. During the spring period of westerly winds, the dune appears to be dominantly in an erosional phase. All west- and north-facing slopes are erosional. Deposition takes place only on the eastern slope of the

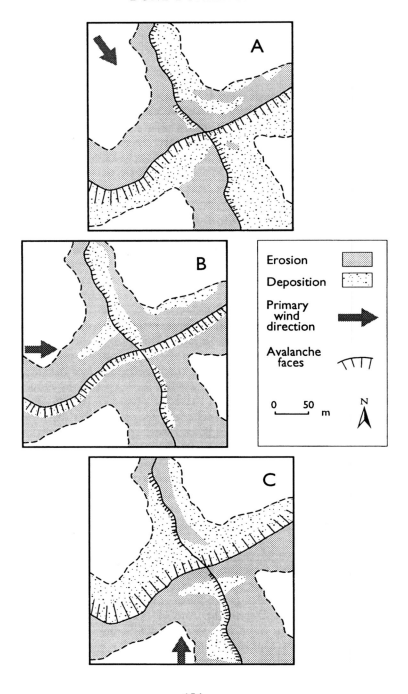

north arm and locally on the southern slopes of the east and west arms. In the summer season of southerly winds, the southern half of the dune and the east slopes of the north arm are eroded, whereas deposition dominates the remainder of the northern half of the dune. In all seasons, most of the deposition occurs on the avalanche faces, but considerable lateral movements of sand also occur.

RELATIONS BETWEEN EROSION AND DEPOSITION PATTERNS AND WINDS

Field observations on all dune types show that there is a very close correspondence between patterns of erosion and deposition and wind velocity and direction. Erosional slopes are those which experience accelerating or divergent winds. Deposition takes place in two settings: rapid and large-volume deposition on lee-side avalanche faces, where flow separation occurs; and slow deposition in areas where sand transport rates decrease downwind as a result of flow convergence or local flow expansion.

The relations between airflow and erosion and deposition patterns can be explained most easily through principles of sediment conservation. Assuming that sand transport rates are proportional to wind velocity (providing that it is above the threshold for sand movement) then spatial changes in sediment transport rates (dq/dx) and thus erosion rates can be approximated by the local amplification factor (Az). Direct relations between the amount of surface change and amplification factors can be demonstrated for linear and star dunes (Figure 5.32). For Namib linear dunes, there is a good correspondence between the measured spatial pattern of erosion and deposition and that computed using wind velocity data (Figure 5.33). However, the model tends to overestimate the amount of erosion and deposition that occurs on the dune. Howard *et al.* (1978), using a similar model, likewise found a close correspondence between measured and simulated patterns of erosion and deposition on a barchan dune.

DEVELOPMENT OF DUNE PROFILES

Field studies have shown that there is a high degree of interaction between the shape of the dune, the amount of change in wind velocities and sand transport rates and erosion and deposition rates. Given that wind velocities increase up the windward slopes of all dunes, by a varying amount and in a pattern determined by the profile of the dune, it follows that all windward slopes tend to be erosional. Downwind, the wind will have to transport an

Figure 5.30 Patterns of erosion and deposition on a Gran Desierto star dune (after Lancaster 1989a). A: winter northerly winds; B: spring westerly winds; C: summer southerly winds.

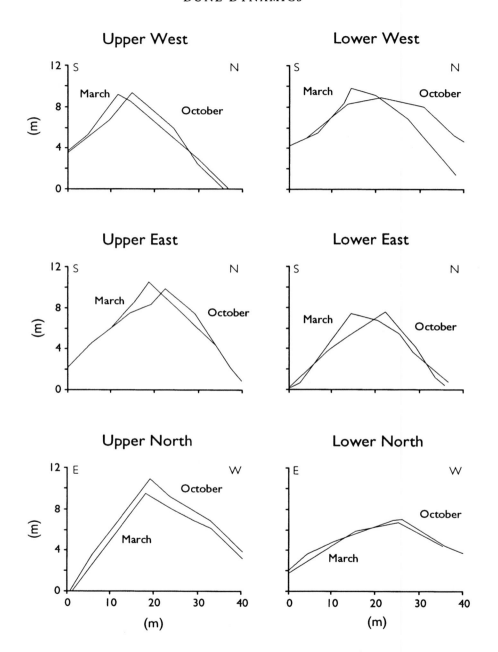

Figure 5.31 Changes in dune profiles on a Gran Desierto star dune
(after Lancaster 1989a).

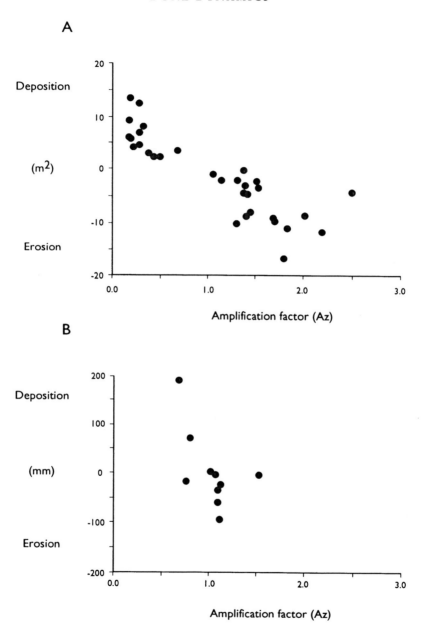

Figure 5.32 Relations between wind velocity amplification and magnitude of erosion and deposition (after Lancaster 1994a). A: Gran Desierto star dunes; B: Namib linear dunes.

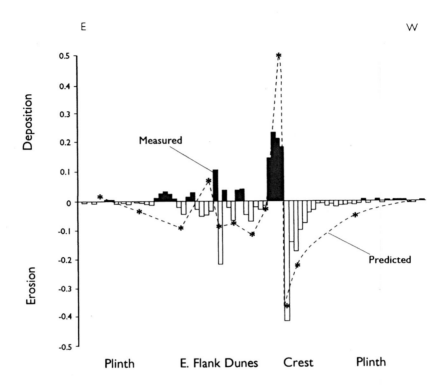

Figure 5.33 Relation between measured and modelled erosion and deposition on a Namib linear dune (after Lancaster 1989b).

increasing amount of sand eroded from the dune slope. This in turn requires that wind velocities and surface shear stress should change to increase transport rates proportionately. If they do not, deposition will occur, leading to adjustment of dune form. The shape of the upwind profiles of dunes thus appears to be adjusted to maintain an equilibrium between the rate of erosion and the increase in velocity required to keep an ever increasing amount of sand in transport. There is thus a high degree of interaction between the shape of the dune, the amount of change in wind velocities and sand transport rates and the rate and pattern of erosion or deposition (Lancaster 1985b; Tsoar 1985; Lancaster 1987).

The shape of the stoss slopes of simple dunes in unidirectional winds can be evaluated using the sediment continuity equation (Fredsoe 1982). Thus:

$$dh/dt = -dq/dx \qquad (5.10)$$

where h is local bed elevation, q is local volumetric sediment transport in the

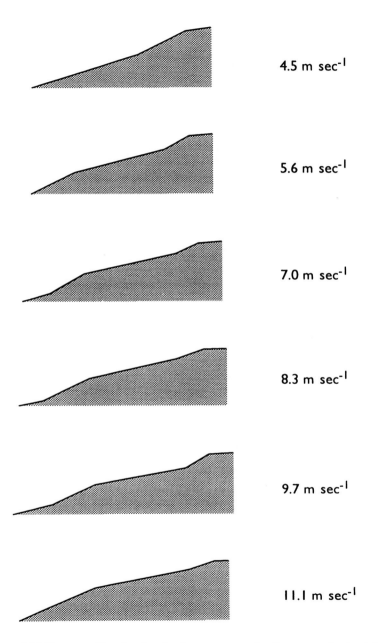

4.5 m sec⁻¹

5.6 m sec⁻¹

7.0 m sec⁻¹

8.3 m sec⁻¹

9.7 m sec⁻¹

11.1 m sec⁻¹

Figure 5.34 Dune profiles simulated using concepts of sediment continuity (after Lancaster 1985b).

direction x and t is time. Each dune is migrating at a rate Q_B given by

$$Q_B = q \text{ (crest)} / H \qquad (5.11)$$

where H is the height of the dune. Combining the two equations (Fredsoe 1982) gives

$$h/H = q \text{ (local)} / q \text{(crest)} \qquad (5.12)$$

Data on velocity amplification can be used to estimate the magnitude of q (local) for any point on the dune via sand transport equations and thus to evaluate the shape of transverse bedforms which will be in equilibrium with a given wind velocity. Simulated dune profiles derived in this way (Figure 5.34) suggest that in periods of low wind velocities dunes tend to develop rather steeper profiles compared to those developed at high velocities. Transverse dune ridges become broader and more rounded as wind velocities increase, in a similar manner to sub-aqueous bedforms (Fredsoe 1982).

As discussed by Tsoar (1985), dunes subjected to bi-directional wind regimes tend to have sharp crestlines and a triangular cross section, rather than the convex form of the crest common to most flow-transverse dunes. The triangular profile is the result of the processes discussed above operating from two directions such that, at a given season, each profile will tend towards a convex form as it adjusts to that wind direction and its spectrum of wind velocities. The high rates of erosion and deposition observed in the crestal areas of linear and star dunes are a reflection of the process of adjustment to changing wind directions.

Dune profiles can therefore be considered to be in a state of dynamic or quasi-equilibrium (Chorley and Kennedy 1971) with prevailing airflow conditions, in which dune form fluctuates about a series of 'average' states, which may change through time as the dune accumulates. The characteristics of the 'average' state may be determined by the magnitude and frequency of sand-moving winds such that there are formative events which determine the overall characteristics of dune form.

CONCLUSIONS

Fundamentally, the initiation, dynamics, and morphology of all dunes are determined by spatial and temporal variations in winds and sediment transport. Recent field studies have provided a wealth of information on winds, sand transport, and erosion and deposition patterns on dunes. All dunes are characterized by flow acceleration on windward (stoss) slopes. This results in an increase in sediment flux towards the dune crest and a net erosion of sediment. The rate of increase of wind shear velocity and sediment flux are controlled by the dune shape (aspect ratio) and by the initial wind velocity conditions at the toe of the slope. Lee side airflow on dunes is often complex and involves a combination of flow expansion, separation, and diversion that lead to deposition and/or along slope transport of sediment. The nature of lee side flow is determined by dune shape and the angle of attack of the wind relative to the crestline. As these variables change in both time and space, the result is a complex series of interactions between form and process that can be documented by studies of erosion and deposition patterns on dunes. These investigations provide information on how dunes behave under different wind conditions and demonstrate the importance of interactions between the dune form and airflow so that dunes tend to evolve towards some form of dynamic equilibrium. Such observations provide the basis for explaining why dunes of different types occur in different wind regimes, as will be discussed in the following chapter.

6

CONTROLS OF DUNE
MORPHOLOGY

INTRODUCTION

A major goal of studies of the geomorphology of desert dunes is to explain
how and why dunes of different morphological types (e.g. crescentic, linear,
star, etc.) and varieties (simple, compound and complex dunes) form, and to
understand the factors that control the size and spacing of dunes. Recent
developments in studies of dune processes and dynamics, combined with
better information on the spatial distribution of dunes of different morphological types derived from satellite images, mean that it is possible to develop
general models for the controls of dune morphology, although many aspects
remain to be fully investigated.

FACTORS THAT MAY DETERMINE DUNE
MORPHOLOGY

Because dunes are a product of interactions between the wind and sand
surfaces, the characteristics of both dune sediments and winds play important roles in determining their type, size, and spacing. These interactions are
modulated by vegetation and the effects of time as dunes grow and interact
with the local wind regime.

Five main groups of factors that influence the type, alignment, size and
spacing of dunes can be identified: (1) the characteristics of the wind regime,
especially its directional variability, have frequently been regarded as the
major factor determining dune type. Wind strength may also be a factor;
(2) sand supply is considered by some workers to be a determinant of dune
type, although satisfactory definitions of this parameter are difficult to
achieve; (3) the nature of dune sands, especially their grain size and sorting
characteristics, has been regarded as a major influence on the size and spacing
of dunes; (4) vegetation cover may be locally important,; and (5) as dunes are
the product of an ongoing process of sediment accumulation, the effects of
time should also be considered.

Dune form and sediment characteristics

Wilson (1972) suggested that different elements of a complex dune pattern are composed of sands with different grain sizes. Grain size, by controlling the threshold velocity for sand movement, may determine the effective wind regime, and therefore strongly influence dune alignments. In areas of complex linear dunes in the Namib Sand Sea, sand from east flank barchanoid dunes and corridor crossing linear dunes is slightly less well sorted and coarser than adjacent complex linear dune crests (Figure 6.1), but the differences are not significant in terms of threshold velocities and therefore effective sand-transporting wind regimes (Lancaster 1989b).

There appears to be no evidence for a genetic relationship between the grain size and sorting character of sands and dune type, except that sand sheets and zibar in many sand seas are composed of coarse, poorly sorted, often bimodal, sands. These relationships have been documented in Saharan sand seas (Bagnold 1941; Capot-Rey 1947; Warren 1972; Maxwell and Haynes 1989) , the Sinai (Tsoar 1978), and in the Skeleton Coast (Lancaster 1982a) and Algodones dune fields (Nielson and Kocurek 1986).

It is not clear how the presence of significant quantities of coarse sand leads to the formation of sand sheets or zibar. Very coarse grains require a high threshold velocity for movement and are therefore moved infrequently by the wind. This relative stability may facilitate the growth of vegetation after occasional rains, which further protects the surface (Kocurek and Nielson 1986) so that in many sand seas, sand sheet areas are often better vegetated

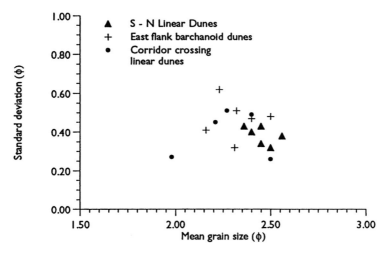

Figure 6.1 Relations between grain size and sorting parameters and dune type in the northern Namib Sand Sea.

than adjacent dunes, and vegetation persists for much longer after rains compared to adjacent dune areas.

Zibar and sand sheets may form low relief surfaces because they cannot trap sufficient fine sand for the development of dunes. However, Nielson and Kocurek (1986) have shown that small-scale flow expansion does take place in the lee of zibar, leading to localized deposition of fine sand. As discussed in Chapter 5, differences in transport rates between the base and crest of bedforms composed of coarse sand can be large (Lancaster 1985b; Tsoar 1986), so that they tend to develop flatter profiles because sand transport is concentrated at the crest. Such bedforms have a low aspect ratio, so that flow in the lee remains attached, and sand can be transported off the form in all but very low wind speed conditions. As a result, growth of the bedform is inhibited.

Dune form and wind regimes

The association of dunes of different morphological types with wind regimes that have different characteristics, especially of directional variability, has been noted by many investigators. Examples of typical wind regimes for different dune types (Chapter 3) show that crescentic, linear, and star dunes are associated with distinctly different wind regimes.

Fryberger (1979) compared the occurrence of each major dune type (crescentic, linear, star) on Landsat images of sand seas with data on local wind regimes. The ratio between resultant (vector sum) drift or sand transport potential (RDP) and total drift potential, DP (potential sand transport from all directions) or RDP/DP, is an index of directional variability equivalent to the unidirectional index (UDI) of Wilson (1971). High RDP/DP ratios characterize near unimodal wind regimes whereas low ratios indicate complex wind regimes. Fryberger (1979) found that the directional variability or complexity of the wind regime increased from environments in which crescentic dunes are found to those where star dunes occur (Figure 6.2). Crescentic dunes occur in areas where RDP/DP ratios exceed 0.50 (mean RDP/DP ratio 0.68) and are frequently found in unimodal wind regimes, often of high or moderate energy. Linear dunes were observed in wind regimes with a much greater degree of directional variability and commonly exist in wide unimodal or bimodal wind regimes with mean RDP/DP ratios of 0.45. The trend of linear dunes is approximately parallel to the resultant direction of sand transport. Star dunes occur in areas of complex wind regimes with RDP/DP ratios less than 0.35 (mean = 0.19).

Comparison of the distribution of dunes of different morphological types in the Namib Sand Sea with the information on sand moving wind regimes (Lancaster 1983a; Lancaster 1989b) confirms Fryberger's hypothesis that there is both an increase in the directional variability of the sand-moving wind regime from crescentic to star dunes as well as a decrease in the overall

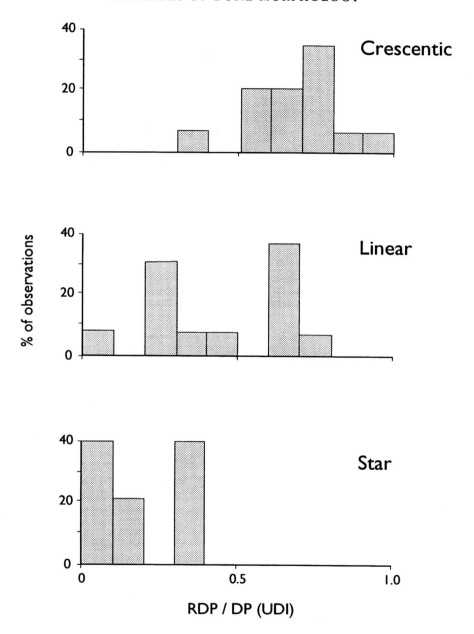

Figure 6.2 Relations between dune type and wind regime in the global sample of dune areas examined by Fryberger (1979).

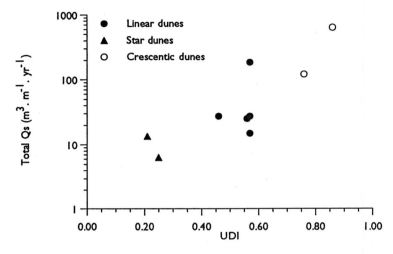

Figure 6.3 Relations between dune type and wind regime (both directional variability as defined by the ratio between total and resultant sand transport potential and total transport potential) in the Namib Sand Sea.

sand transport potential (Figure 6.3). Thus star dunes in the Namib occur in low energy, complex wind regimes, and barchans in high energy, unimodal wind regimes. However, the overall directional variability of wind regimes in the Namib Sand Sea is less than that in Fryberger's global sample.

Studies of dune processes and dynamics help to explain why dunes of different types occur in different wind regimes. The primary response of sand surfaces to the wind is to form an asymmetric dune with a convex stoss slope and a steep lee slope (Tsoar 1985). The strike of the crestline is approximately perpendicular to the mean wind direction. In the form of crescentic dunes of different varieties, this is the most common dune type in unidirectional wind regimes. It also characterizes small dunes in many multidirectional wind regimes. In multidirectional wind regimes, profiles of the crestal areas of linear and star dunes tend towards this form in each wind season (see Chapter 5). As dunes grow, their cross-sectional area increases exponentially (Figure 6.4). As a result, they can no longer be remodelled in each wind season, so form–flow interactions become significant, and they develop a morphology that is controlled by more than one wind direction. There are many examples of dunes of different types occurring together in the same wind regime (e.g. Nielson and Kocurek 1987; Lancaster 1989a): the small dunes are almost always crescentic forms because they can be reformed entirely in each wind season; larger dunes tend to be linear or star types. Star, and possibly linear dunes, are therefore not primary dune types: they develop as a result of the

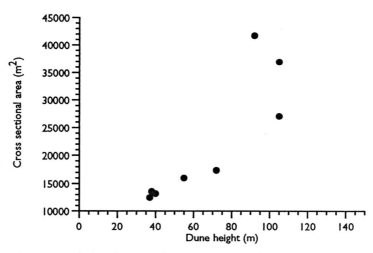

Figure 6.4 Relations between dune cross-sectional area and dune height.

modification of other dunes as they migrate into areas of different wind regimes (Tsoar 1974; Lancaster 1989a).

Development of linear dunes by modification of barchans was first suggested by Bagnold (1941) by reference to a hypothetical situation in which a barchan dune moved into a bi-directional wind regime. Over time, it would be transformed into a linear dune by the elongation of one horn (Figure 6.5a). Strong winds from an oblique direction would add sand to one horn which would then be extended by gentler winds from the original direction. Repetition of this cycle would lead to the creation of a linear dune with regularly repeated summits and a sharp sinuous crestline. Support for this model comes from the Namib where Lancaster (1980) reports linear dunes 1–1.5 km long and 3–8 m high with a WSW–ENE alignment developing from barchans aligned transverse to WSW winds. Lancaster argued that strong southwesterly and south-southwesterly winds supply sand to the linear dunes which is then redistributed along the dunes by weaker west-southwesterly to west-northwesterly winds. Tsoar (1984) suggested a different model from observations in the Sinai, where winds from both the primary and secondary directions extend the horn on the side of the dune opposite to the secondary wind direction (Figure 6.5b). The linear element of the dune extends more rapidly than the original barchan can migrate, and the dune evolves to a linear form.

Process studies have demonstrated that the basic mechanism for linear dune formation is the diversion of winds approaching at an oblique angle to the crest to flow parallel to the lee side and transport sand along the dune (Tsoar 1983a). Thus any winds from a 180° sector centred on the dune will

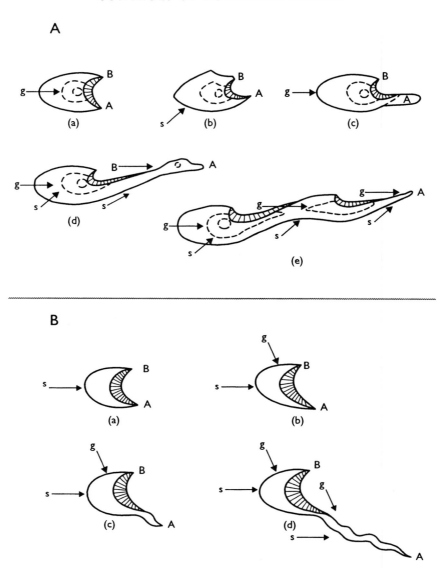

Figure 6.5 Models for the formation of linear dunes by modification of barchans. A: after Bagnold 1941; B: after Tsoar 1984. *s* = strong winds, *g* = gentle winds; *A* and *B* indicate barchan horn positions for reference.

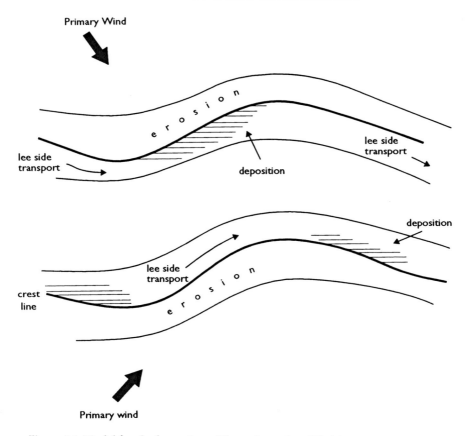

Figure 6.6 Model for the formation of linear dunes (modified from Tsoar, 1983a).

be diverted in this manner and cause the dune to elongate downwind (Figure 6.6). This will not necessarily take place in a direction parallel to the resultant direction of sand transport, but most probably at an angle of 20–30° to the most effective sand-transporting wind direction (Tsoar 1983a). Linear dunes in the Sinai extend at an angle of around 40° to northwesterly summer winds. The overall trend of complex linear dunes in the Namib Sand Sea is determined by the dominant SSW to SW winds and the dunes extend at an angle of 20–30° to this directional sector (Livingstone 1988; Lancaster 1989b). Sand is moved along the dune by the dominant SSW winds, whereas strong easterly winds rework dune crests and deposit sand on the dune (Figure 6.7)

Winds from two directions are essential for the maintenance of the linear form. Linear dunes are not stable in a unidirectional wind regime because

171

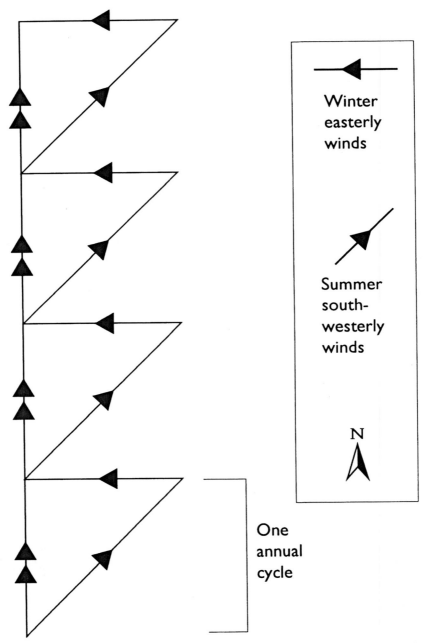

Winter
easterly
winds

Summer
south-
westerly
winds

N

One
annual
cycle

Figure 6.7 Model for the formation of linear dunes (after Livingstone 1988).

erosion and deposition are concentrated at the same places on the dune, resulting in their eventual breakup (Tsoar 1983a). Thus a seasonal change in wind direction is necessary to maintain the dune and its triangular profile. In bimodal wind regimes where the modes are 180° or less apart, deflection of all oblique winds on the lee side will tend to elongate the dune. The effects of each wind direction will be determined by their direction relative to the dune. If a high percentage of winds are at optimal angles for lee side deflection then the dune will extend strongly, and such dunes will tend to be long and relatively low. If winds blow at higher angles to the dune, then movement of sand along the dune will be replaced by deposition on lee side avalanche faces. Sand will therefore tend to stay on the dune and increase its height. A limiting case will occur when winds are perpendicular to the crest, producing a reversing dune. Such dunes tend to accrete vertically, and develop major lee side secondary flow cells that move sand toward the centre of the dune, leading to the formation of star dunes.

The development of star dunes is strongly influenced by the high degree of form–flow interaction that occurs as a result of seasonal changes in wind direction, and the existence of a major lee-side secondary circulation (Lancaster 1989a). This results in the reworking of deposits laid down in the previous wind season. Sand, once transported to the dune, tends to stay there and add to its bulk. This is a result of the high deposition rates on lee faces and the fact that wind velocity and sand transport rates decrease away from the central part of the dune. The prominent lee-side secondary flows also tend to move sand towards the centre of the dune and promote the development of the arms.

Observations of dune morphology in several sand seas suggest that star dunes most probably form by the modification of pre-existing dunes as they migrate or extend into areas of opposing or multidirectional wind regimes, along the lines illustrated schematically in Figure 6.8. Modification of linear dunes to chains of star dunes is reported from the Namib Sand Sea by Lancaster (1983a). Observations of sand seas in Arabia and north Africa, and at Great Sand Dunes, Colorado, show that in these areas, star dunes lie down the net sand transport direction from areas of crescentic dunes (Breed et al. 1979; Andrews 1981). Nielson and Kocurek (1987) have suggested that star dunes at Dumont, California, may be initiated by the migration and merging of crescentic dunes in a seasonally variable wind regime. They observed that small star-like dunes formed during periods when winds changed direction seasonally, but were remodelled to barchans in the constant direction winds of each season, and suggested that there was a minimum 'survival size' for star dunes. Once initiated, the star dunes are maintained by secondary airflow and winds from multiple directions. This hypothesis is supported by observations of dunes in the Gran Desierto.

In the eastern part of the Gran Desierto, simple and compound crescentic dunes migrate towards the north (Lancaster 1989a). As they do so, they enter

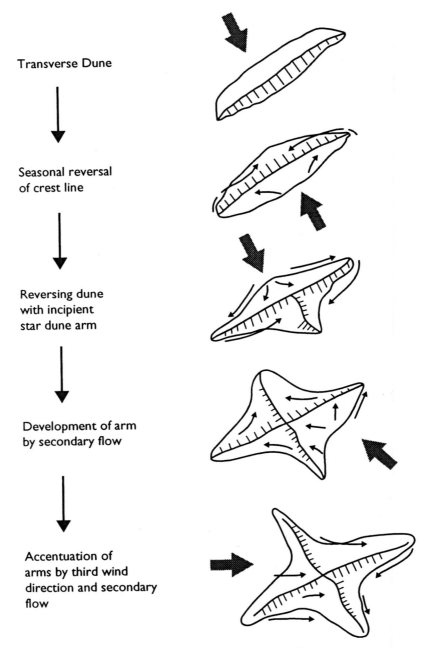

Transverse Dune

Seasonal reversal
of crest line

Reversing dune
with incipient
star dune arm

Development of arm
by secondary flow

Accentuation of
arms by third wind
direction and secondary
flow

Figure 6.8 Model for the formation of star dunes (after Lancaster 1989a).

a region where an increased frequency or strength of northerly winds occurs. The crescentic dunes first develop reversing dunes on their crests, and by the northern margin of the sand sea large star dune peaks have developed on top of these dunes. In the northwestern part of the sand sea, crescentic dunes migrate towards the southeast. Such dunes encounter opposing southerly winds which reduce their rates of forward migration, such that faster moving dunes situated upwind eventually collide with them, coalesce and add to their bulk. Deposition in such locations will tend to be self reinforcing, and lead to the growth of first reversing and then star dunes, a sequence that can be observed in many parts of the western Gran Desierto.

Sand supply

The availability of sand for dune building has long been considered a factor influencing dune morphology (e.g. Hack 1941; Wilson 1972). Using a sample of dunes of different types from sand seas in all major desert areas, Wasson and Hyde (1983b) established that the mean EST (equivalent or spread-out thickness of dune sand in a given area) for all dune types was statistically identical and concluded that although sediment availability was a significant variable determining dune type, it was not the only one. However, by plotting EST against Fryberger's RDP/DP ratio, a clear discrimination of dune types was achieved (Figure 6.9), leading to the conclusion that barchans

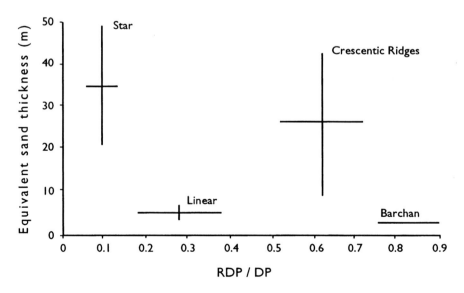

Figure 6.9 Relations between dune type, wind regime, and sand supply (reprinted in a modified form from Wasson and Hyde 1983b, with permission from *Nature*).

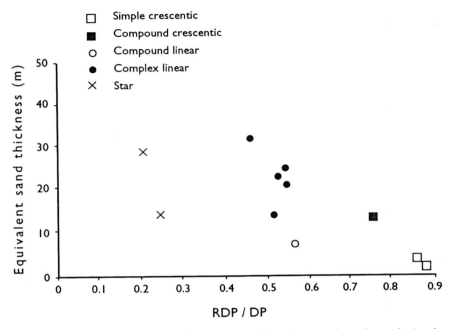

Figure 6.10 Relations between dune type, wind regime, and sand supply in the Namib Sand Sea (after Lancaster 1989b).

occur where there is very little sand and almost unidirectional winds: crescentic dunes where sand is abundant and winds slightly more variable; linear dunes occur where sand supply is small, but winds more variable still; and star dunes are found in complex wind regimes with abundant sand supply.

Similar relations are evident in the Namib Sand Sea, although the range of directional variability is less (Figure 6.10). All dune types in the Namib Sand Sea occur in wind regimes which are less variable in direction than Wasson and Hyde's sample. This is especially true for linear and crescentic dunes. In addition, star dunes in the Namib Sand Sea occur where EST is lower than the Wasson and Hyde model, and linear dunes, especially of complex varieties, where EST is higher than predicted. In the Namib Sand Sea there is a increase in EST as wind regimes become more variable (Lancaster 1989b).

However, EST is not a measure of sand supply but of the volume of sand contained in the dunes, and may be a reflection of dune type with the dune type being influenced by other factors, especially the wind regime (Rubin 1984). In the Namib Sand Sea, it is possible to discriminate clearly between dune types on the basis of their relations to wind regimes. The EST data merely suggests that there is more sand in complex linear and star dunes than in compound linear and all types of crescentic dunes.

One situation where sand availability may be important is the development of sand sheets. Sand sheets in the northwestern parts of the Gran Desierto are composed of sands that are in the range commonly found in dunes, and their formation is not the result of a significant proportion of coarse or very coarse sand which acts to suppress the formation of dunes with slip faces (Lancaster 1993a). The modal size fraction in the Gran Desierto sand sheets is 180 to 250 μm compared to 350 to 710 μm in Namibian and Algodones sand sheets and zibars. The Gran Desierto sand sheets are also unimodal with a prominent mode at 250 μm in contrast to many unvegetated sand sheets and zibars, which tend to be bimodal (Nielson and Kocurek 1986). Today, the vegetation cover in the area is sparse, with a maximum percentage cover of 10–15 per cent. This is insufficient to prevent sand transport taking place, but sufficient to cause divergence and convergence of airflow around individual plants in the manner suggested by Fryberger et al. (1979) and Ash and Wasson (1983) and give rise to localized deposition. If sand supply was high then it is possible that these effects could lead to dune initiation (Kocurek et al. 1992). However, dunes are not being initiated by these processes in this part of the Gran Desierto. It appears therefore that sand supply is a limiting factor, and the sparse vegetation cover leads to deposition of a poorly sorted mixture of fine and coarse sand. Many sand sheets occur on the upwind margins of dune areas and in interdune areas between star and linear dunes. They may therefore be the product of a limited sand supply and/or a wind regime that promotes sediment bypassing. Sediment bypassing occurs when winds and sand transport rates do not change in the direction of transport.

Vegetation

The effects of vegetation on sand transport rates and nucleation of dunes have been discussed in earlier chapters. Sand is an excellent medium for plant growth in deserts, because of its high moisture holding capacity (Bowers 1982). Many dunes, even in hyper-arid regions like the Namib, are vegetated to some extent, mostly in the plinth areas of linear and star dunes where relatively little surface change occurs (Thomas and Tsoar 1990).

The direct effects of vegetation on dune form are not well understood. Dunes that migrate actively, such as barchans and many crescentic ridges, tend to be less vegetated than linear and star dunes in the same climatic zone (Thomas and Tsoar 1990), because the processes of stoss slope erosion and lee deposition create an environment that is unfavourable for plant growth. Parabolic dunes and shadow dunes are two examples of dune types that evolve in the presence of vegetation (Hesp 1981; Wasson et al. 1983). Hack (1941) suggested that in northeastern Arizona there was a transition from

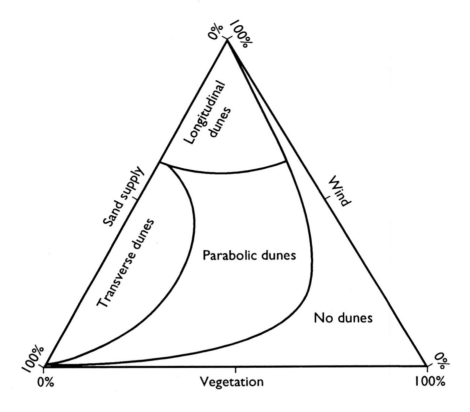

Figure 6.11 Relations between sand availability, winds, and vegetation cover (after Hack 1941).

crescentic to parabolic dunes with increased vegetation cover, and linear dunes occurred in areas with less sand and vegetation than parabolic dunes (Figure 6.11). Tsoar (1989) has distinguished between straight 'vegetated linear dunes' that are widespread in Australia and the Kalahari and sinuous crested 'seif' type linear dunes. He argues that straight linear dunes form in the presence of a significant vegetation cover. They extend parallel to the strongest sand-moving wind direction, as the vegetation cover on their flanks restricts the effectiveness of weaker winds via increases in aerodynamic roughness. When the vegetation cover is reduced or eliminated, vegetated linear dunes may evolve toward sinuous 'seif' types of linear dunes as all wind directions can become effective in sand transport (Tsoar and Møller 1986). However, stratigraphic evidence (e.g. Heine 1982; Gardner *et al.* 1987; Nanson *et al.* 1992) shows that many linear dunes of the type widespread in

Australia and the Kalahari were originally formed during the late Pleistocene and have since become stabilized by vegetation in more humid and/or less windy conditions. The dune patterns are therefore not the result of current processes and the dune morphology observed today is the result of the degradation of a former sinuous, sharp crested form. Where dunes are still active in these regions (or have been reactivated by vegetation disturbance) their form is similar to that of other unvegetated linear dunes with a sinuous sharp crestline.

CONTROLS ON DUNE SIZE AND SPACING

Most sand seas show clear patterns of dune size and spacing, and the close correlations that exist between height, width and spacing suggest a high degree of adjustment to controlling variables. However, the nature of the controls on dune size and spacing are not well understood.

Relations between dune size and spacing

A range of sizes of aeolian bedforms occurs in all desert sand seas. As can be seen from Figure 6.12, the dune height to spacing ratio varies widely between dune types and areas. The ratio for crescentic dunes clusters around 0.04, with a range from 0.02–0.08; whereas that for linear dunes ranges between 0.01 and 0.05, with a mean of 0.04. In areas of crescentic dunes, the ratio appears to increase with sand supply (Kocurek *et al.* 1992), and at Padre Island, the dune pattern evolves to a height–spacing ratio of 0.05. The upper limit for the dune height to spacing ratio is apparently set by aerodynamic constraints that indicate a minimum possible dune spacing of approximately 4 dune heights or the width of the lee side separation zone (Sweet 1989). The lower limit may be a function of sand availability.

The relations between dune height and spacing are a reflection of the amount of sand incorporated in the bedforms of an area (Lancaster 1988a), as well as the way in which the system is self-organized (Hallet 1990). The general form of the relations between dune height and spacing can be expressed by a power function

$$D_H = c D_S^n \qquad (6.1)$$

where D_H is dune height, D_S is dune spacing, c is a constant and the exponent n is a measure of the rate of change of the dependent variable relative to the rate of change of the independent variable (the slope of the regression line). Values of the exponent n range between 0.52 and 1.72 from one sand sea to another as well as from one dune type to another in the same sand sea. An exponent of unity indicates that dune height increases at the same rate relative to dune spacing. Thus a given amount of sand can be formed into a few widely spaced dunes or many small closely spaced dunes. Such a model was

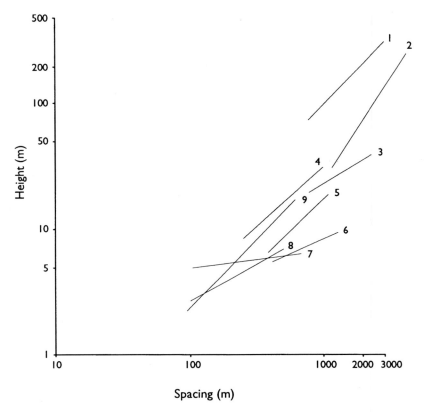

Figure 6.12 Relations between dune height and spacing in different sand seas (after Lancaster 1988a). (1) Namib Sand Sea star dunes, (2) Namib Sand Sea complex linear dunes, (3) Namib Sand Sea compound linear dunes, (4) Namib Sand Sea crescentic dunes, (5) Simpson-Strzelecki simple linear dunes, (6) Great Sandy Desert linear dunes, (7) Skeleton Coast crescentic dunes, (8) Gran Desierto crescentic dunes, (9) Southwestern Kalahari linear dunes.

advocated for Simpson Desert linear dunes by Twidale (1972). Where the exponent is greater than unity (e.g. complex linear dunes in the Namib Sand Sea and star dunes in the Gran Desierto), dune height increases more rapidly than dune spacing, indicating a tendency for vertical growth of the dunes. Exponents less than unity indicate that dune height increases less rapidly than dune spacing. In these examples (compound linear dunes in the Namib Sand Sea, linear dunes in the Great Sandy Desert, crescentic dunes in the Gran Desierto) dune size may be limited by the availability of sand. Relatively small dunes for their spacing have also been noted from areas undergoing net sand loss (Mainguet and Chemin 1983), and from areas of damp interdunes at Padre Island (Kocurek *et al.* 1992). The relation between dune height and spacing therefore appears to reflect both the availability of sand for dune building and wind regime characteristics, which determine whether dunes will tend to accrete vertically (star dunes and complex dune varieties), migrate (simple crescentic dunes), or extend (many simple and compound linear dunes).

Grain size effects

Wilson (1972) found a clear relation between dune and draa (compound and complex dune) spacing and .the grain size of the coarse twentieth percentile of dune crest sands (P_{20}) in three northern Saharan sand seas (Figure 6.13). He argued that the spacing of dunes and draas was related to the size of secondary flow elements which were in turn related to the wind shear velocity required to transport the sand from which they were composed. For a given grain size, there is a minimum threshold shear velocity which will move this sand and hence a minimum possible dune spacing. The grain size of P_{20}, according to Wilson (1972), closely approximates the mean size of surface sand in mixed deposits. In this model, changes in dune spacing are therefore caused primarily by changes in the grain size of their sediments.

However, data for dune fields and sand seas in the Namib, the southwestern Kalahari, and the Gran Desierto of Mexico (Figure 6.14) show that there is no general relation between dune spacing and grain size in these areas (Lancaster 1988a). Similarly, recent data from Australia and the Kalahari (Wasson and Hyde 1983a; Thomas 1988) show no relation between P_{20} and linear dune spacing. These data suggest that Wilson's hypothesis of a grain size control of dune spacing is not generally applicable. In some sand seas, the spacing of some dune types is apparently correlated with P_{20}, whereas

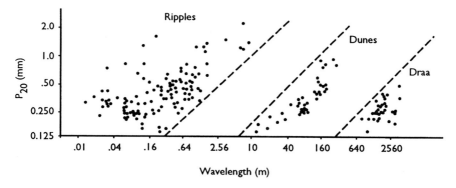

Figure 6.13 Relations between dune spacing and P_{20} in Saharan sand seas (after Wilson 1972).

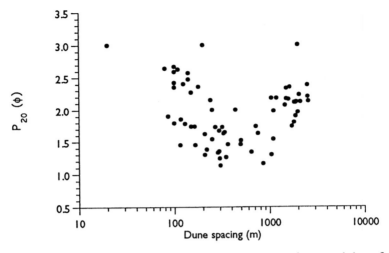

Figure 6.14 Relations between dune spacing and P_{20} in sand seas and dune fields in the Namib, Kalahari, and Gran Desierto (after Lancaster 1988a).

other types show no relation (Figure 6.15). The spacing of complex linear and star dunes in the Namib and Gran Desierto sand seas is unrelated to P_{20}. However, data for simple and compound crescentic dunes in both areas show that their spacing tends to increase with a coarsening of P_{20} (Figure 6.15a).

One possible mechanism to explain the spacing of crescentic dunes in terms of grain size and changes in winds and transport rates over dunes was put forward by Lancaster (1985b), who suggested that dunes of a given height composed of fine sands will tend to develop shorter, steeper stoss slopes compared to those composed of coarse sand (Figure 6.16). Coarse sands may therefore result in broad dunes with a large crest-to-crest spacing for their height, as also observed by Tsoar (1986). This model may be a partial explanation of the observed relationships between crescentic dune spacing and P_{20}.

Effects of wind regimes

In sub-aqueous environments, bedform spacing increases with flow depth and grain size (Allen 1968b). If depth and grain size are held constant, then bedform size scales with flow strength and thus transport rates (Kennedy 1969). Downstream changes in transport rates therefore lead to a decrease in

Figure 6.15 Relations between dune spacing and P_{20} for crescentic, linear, and star dunes.

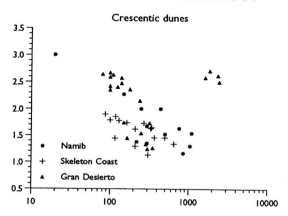

Crescentic dunes

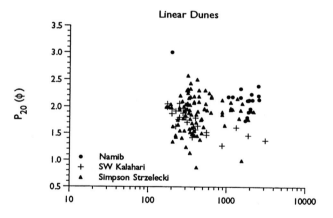

Linear Dunes

$P_{20} (\phi)$

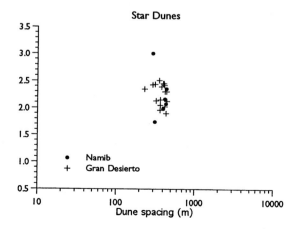

Star Dunes

Dune spacing (m)

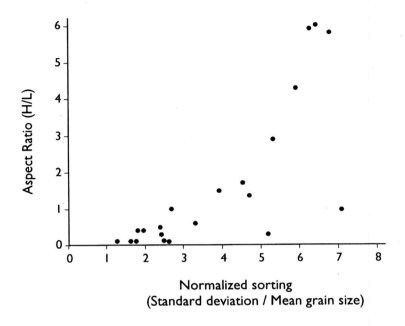

Figure 6.16 Changes in profile form with grain size (after Tsoar 1986).

bedform size and/or the number of bedforms (Rubin and Hunter 1982).

By analogy with sub-aqueous bedforms, it might be expected that dune size is related to mean sand transport rates, so that larger dunes are located where sand transport rates are high, and small dunes occur where transport rates are low (Wilson 1972). However, for 100 km² areas of dunes in the Namib Sand Sea the opposite situation occurs: large dunes are found in areas where annual potential sand transport rates are low and small dunes occur in areas of high potential sand transport rates (Figures 6.3, 6.17a). Similar patterns appear in Saharan sand seas, where large dunes are associated with low total and net sand transport potential (Wilson 1973; Breed *et al.* 1979).

However, when individual complex dunes in the Namib Sand Sea are considered, the height of superimposed crescentic dunes increases with measured wind speeds and calculated sand transport rates at different points on the dune (Figure 6.17b). Similar relations between superimposed dune height and wind speed changes have also been observed by Rubin and Hunter (1982) and Havholm and Kocurek (1988) on compound crescentic dunes, suggesting that superimposed and primary dunes are formed by two different scales of airflow variability and can coexist in the same flow in a similar manner to the sub-aqueous bedforms studied by Smith and McLean (1977) and Rubin and McCulloch (1980).

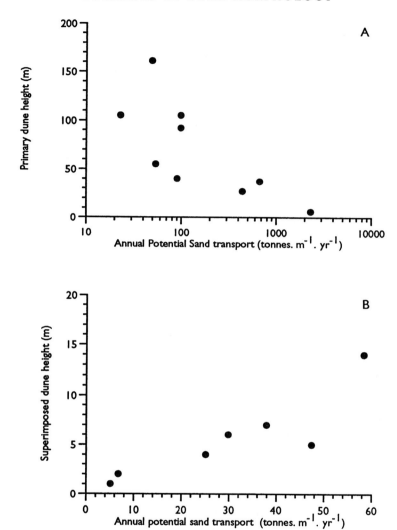

Figure 6.17 Relations between dune height and potential sediment transport in the Namib Sand Sea. A: primary dunes, B: superimposed dunes on flanks of complex linear dunes (after Lancaster 1988a).

AERODYNAMIC MODELS FOR DUNE SPACING

Many investigators have suggested that the spacing of dunes is related to the scale of secondary flow circulations or to the dimensions of the zone of flow disturbance downwind of dunes (e.g. Hanna 1969; Folk 1971; Wilson 1972; Twidale 1972; Tsoar 1978; Lancaster 1983b). Others, e.g. Yalin (1972) and

Cooke and Warren (1973), suggest that the scale of natural atmospheric turbulence is a major influence on dune size and spacing. Howard *et al.* (1978) have argued that possible controls of dune size include upwind roughness, with dune size increasing with the size of fixed roughness elements such as rocks or vegetation.

Helical roll vortex models for linear dune spacing

The parallelism and regular spacing of many linear dune systems has been noted by many workers and has given rise to hypotheses that their spacing is the product of organized vortex flow in the planetary boundary layer (e.g. Folk 1970; Wilson 1972; Tseo 1993). There is a strong tendency for sheared flow in thermally unstable conditions to develop a series of large-scale roll vortices (Brown 1980).

Bagnold (1953b) suggested that the resulting helical motion of the air would impart an alternate left and right hand oblique component to the surface wind which would then sweep sand into long parallel ridges spaced the width of a vortex pair apart (Figure 6.18). In support of this view, Bagnold stated that linear dunes in Egypt were parallel to the resultant of

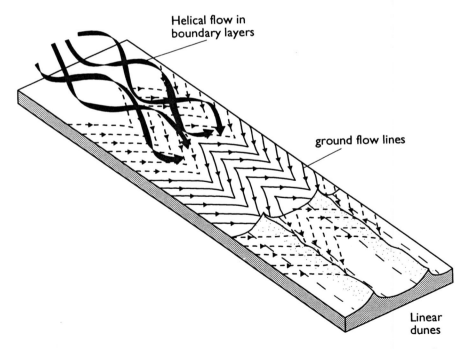

Figure 6.18 Formation of regularly spaced linear dunes by helical roll vortices (after Cooke and Warren 1973).

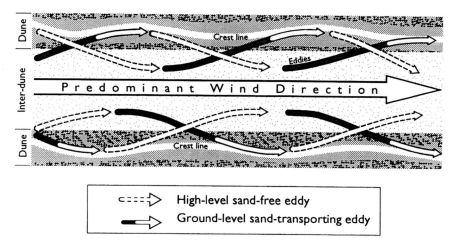

Figure 6.19 Formation of linear dunes by local pressure gradients (after Glennie 1970).

winds during the hottest months but not to those of the whole year. Glennie (1970) argued that pressure gradients between interdune and dune areas would also impart an overall spiral motion to airflow in which surface eddies were directed outwards towards the dunes (Figure 6.19). Tseo (1990) detected evidence for such pressure gradients in the Simpson Desert.

Hanna (1969) compared data on the spacing of linear dunes with the expected dimensions of helical roll vortices in subtropical desert areas, concluding that their probable wavelength was 2–6 km (two to three times the convective layer thickness of 1–2 km). Measured dimensions of roll vortices indicate that their wavelength ranges between less than 50 m and 15 km, with a mode clustering around 5 km (Kelley 1984). However, the maximum spacing of linear dunes in regular patterns is 3–3.5 km (Breed and Grow 1979) and there are large areas, such as the Simpson and Kalahari deserts, where dune spacings are commonly less than 1 km, much smaller than any reported dimensions of helical roll vortices (e.g. Le Mone 1973; Kelley 1984). The observed scale of helical roll vortices is in many cases much larger than average dune spacing, suggesting that linear dunes are not the product of such atmospheric motions. Further, although commonly identified from meso-scale boundary layer studies, evidence for the existence of helical roll vortices in linear dune landscapes is slight. Tseo (1990) interpreted the configuration of tethered kites during periods of dune parallel winds in the Simpson Desert as evidence of small scale (10 m) helical roll vortices, but doubted that they could be responsible for the larger scale ordering of the dune field. Horizontal wind velocities deduced from the tetroon flights of Angell *et al.* (1968) are between 2 and 6 km.hr^{-1} and unlikely, therefore, to move much sand. Angell *et al.* (1968) also noted that helical roll vortices tended to drift laterally to the mean wind. If this is indeed the case, then a stable linear dune system is unlikely to be developed by the action of helical roll vortices (Greeley and Iversen 1985).

The helical roll vortex model for linear dune spacing is attractive, but there is little empirical evidence to support it in dune areas. An alternative model argues that the spacing of linear dunes is related to the distance over which airflow was disturbed in their lee. Tsoar (1978) suggested that the spacing of linear dunes was approximately 12–15 times their height. However, formative winds cross these dunes at an oblique angle and their spacing is much greater than this parallel to the wind. In the Namib Sand Sea, the distance between linear dunes along the line of the most persistent sand moving winds (SSW) averages 23 times dune height (Lancaster 1983b). Dune spacing in areas of simple linear dunes is frequently of the order of 50 times dune height. Observations of the zone of flow separation in the lee of linear dunes (see Chapter 5) suggest that it is confined to a narrow zone immediately adjacent to the dune crest and that flow across interdune areas is constant, but probably undersaturated. Simple wake models such as these are inadequate to explain the spacing of linear dunes.

Models for star dune spacing

Cornish (1914) proposed that star dunes formed at the centres of convection cells, such that winds blowing towards the rising column of air would sweep sand in from the surrounding area. Clos-Arceduc (1966) suggested that star dunes formed at the nodes of stationary waves in oscillating flows. Dunes of star form frequently occur in areas with wind regimes which are multi-directional, or complex, especially in the months in which most sand transport occurs. Sand transport into an area is greater than that from it, leading to the growth of large pyramidally shaped dunes. Allen (1984) has suggested that the spacing of star dunes may be related to the average excursion of sand grains over the duration of the sand-moving winds from each major direction, such that deposition tends to occur at a number of equally spaced nodes. Such nodes would then be self perpetuating and grow to form star dunes. Wilson (1972) proposed that star dunes may develop at the nodal points of complex dune alignment patterns created by crossing or converging sand transport paths. Tseo (1993) has tested a model developed by Huthnance (1982) for the development of linear sand banks in oscillating tidal flows. Although problems of scaling restrict the direct application of the Huthnance model to sand dunes, the concept of linear and star dunes developing at the convergent points of seasonally changing transport pathways is an attractive general model that deserves further exploration.

Crescentic dune spacing

A number of investigators have suggested that the spacing of crescentic ridges is related to the width of the zone in which winds are perturbed in their lee. Cornish (1897) first proposed that a large eddy fills the interdune areas between transverse dunes with flow reattachment on the next dune downwind. Folk (1976b) suggested that pre-existing transverse vortices control the spacing and morphology of crescentic dunes. Studies of winds downstream of natural and artificial obstacles suggest that wind velocities are reduced for a distance up to 12–15 times the obstacle height (Chepil and Woodruff 1963; Oke 1978). In the Sahara, Warren and Knott (1983) noted that turbulent intensity remained high for distances of 300–400 m downwind of small barchan dunes.

However, as noted in Chapter 5, field observations of winds in the lee of crescentic dunes by Sweet and Kocurek (1990) and Frank (in press) show that the maximum width of the separation zone is four dune heights. This means that dune spacing is not a simple function of the size of the lee side eddy. Numerical models and field observations of sub-aqueous bedforms (McLean and Smith 1986) indicate that the redevelopment of an attached boundary layer in the lee of dunes may be a major control of dune spacing. At the point of flow re-attachment, wind speed and shear stress (and hence sand transport rates) are at a minimum. Downwind from this point, an internal boundary layer develops below the wake region created by flow separation at the crest (Figure 6.20). As this internal boundary layer develops, it adjusts to the velocity of the decaying wake region above it. Initially, the wake decays rapidly, causing a large acceleration in the near-bed flow and a rapid increase in surface shear stress. Further downwind, the effects of boundary layer growth overtake the wake effect and decelerate the near surface flow, decreasing the stress. Modifications of this model by Nelson and Smith

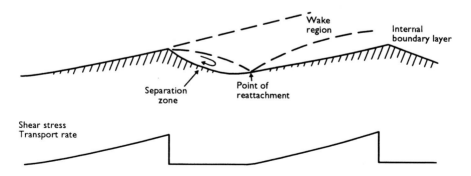

Figure 6.20 Model for the development of a wake downstream of a crescentic dune (after McLean and Smith 1986).

189

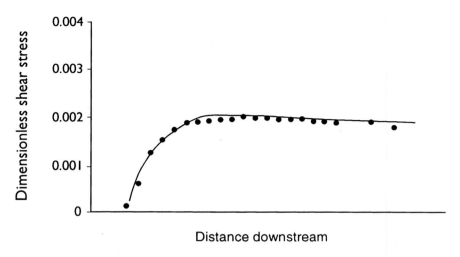

Figure 6.21 Shear stress patterns downstream of an isolated dune (after McLean and Smith 1986).

(1989) include the effects of 'stacked' wakes arising from the existence of other bedforms upstream. In the absence of another dune downwind, the shear stress at the bed, and thus transport rates, increases asymptotically to an equilibrium value downwind of the re-attachment zone (Figure 6.21). McLean and Smith (1986) suggest that the point at which the shear stress maximum is reached defines the position of the next bedform downwind, and is 30–50 times the height of the dune. Following requirements of sediment continuity, then erosion will occur upstream of the shear stress maximum, and deposition downstream. Thus, a new dune will appear and grow at this point until it is large enough to cause flow separation, beginning the process again. If this model is correct, and comparisons with field data suggest that it does represent flow over sub-aqueous bedforms accurately, then the spacing of crescentic dunes may be determined by some combination of average wind speed conditions and overall bedform size (and hence sand availability).

DEVELOPMENT OF COMPOUND AND COMPLEX DUNES

Large dunes are almost always compound and complex forms that are characterized by the development of superimposed smaller dunes upon them. In sub-aqueous environments, there are two major models for the development of superimposed bedforms. Allen (1968b, 1984) and others

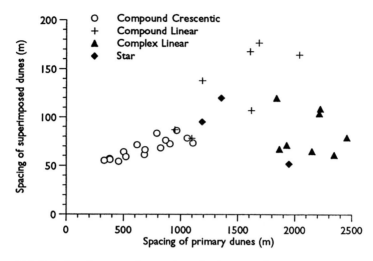

Figure 6.22 Relations between the spacing of primary and superimposed dunes in the Namib Sand Sea (after Lancaster 1988a).

have suggested that they develop in response to temporal changes in flow conditions. For dune areas in Arabia and the Namib, Glennie (1970) and Besler (1980) have suggested that compound and complex dunes are a product of Quaternary climatic changes, in which the large dunes were formed during periods of strong winds during Pleistocene glacial periods. Weaker modern winds can only form the small dunes that are superimposed on their flanks. Support from such a hypothesis comes from complex linear dunes in the Akchar Sand Sea of Mauritania. These dunes are composite forms, with a core formed in the last Glacial period, and superimposed crescentic dunes that are Holocene to Modern in age (Kocurek *et al.* 1991).

Alternatively, Smith and McLean (1977) and Rubin and McCulloch (1980) have shown that two or more scales of bedforms may coexist in equilibrium with steady flow conditions in rivers and estuaries. The widespread existence of compound dunes in clearly active modern sand seas, as well as in the rock record, suggests that superimposed bedforms are the product of contemporaneous aeolian environments.

In the Namib Sand Sea, there is a strong correlation ($r = 0.84$ and 0.75) between the mean spacing of superimposed dunes and that of the draas on which they are situated in areas of compound crescentic and linear dunes, but no such correspondence in areas of complex linear and star dunes (Figure 6.22). This suggests that the size of the superimposed dunes in areas of compound dunes scales with that of the major dune and is the product of two components of the same airflow pattern. In areas of complex linear and star

191

Linear Dunes (n=39)

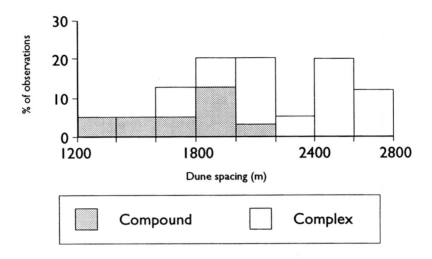

Crescentic Dunes (n=15)

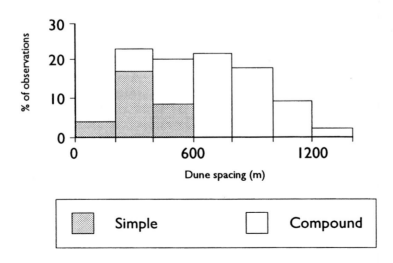

Figure 6.23 The continuum of dune spacings between simple, compound and complex varieties (after Lancaster 1988c).

dunes, there is no relationship between the spacing of the draas and the superimposed dunes, which are largely transverse to sand-transporting winds. Two different types of airflow pattern are apparently involved in this case.

Aeolian bedforms are a response to the dynamic instability of sand transport in conditions of fluctuating flow on planar sand surfaces. Given sufficient sand supply, ripples and dunes will be formed on flat desert surfaces. In the same way, the slopes of large dunes will present an effectively planar surface on which sand transport takes place. Therefore, variations in sand transport rates on compound or complex dunes in time or space (Lancaster 1985b) will lead to the formation of superimposed dunes if the major dune is sufficiently large. This suggests that there is minimum size for compound and complex dunes.

Wilson (1972) argued that dunes and draas (compound and complex dunes) were distinct bedforms, and that the discontinuity in spacing indicated that dunes could not grow into draas. Although the sample size is small, data on the width and spacing of simple, compound and complex crescentic and linear dunes (Breed and Grow 1979) indicate that the mean size of simple, compound and complex dunes is statistically significantly different (Lancaster 1988c). This suggests that a minimum dune size must be reached before superimposed dunes can develop. In the Namib Sand Sea, simple crescentic dunes have a spacing of less than 500 m, whereas compound dunes are all more than 500 m apart. In both the Namib and Australian sand seas (Wasson and Hyde 1983a; Lancaster 1988c) there is, however, a continuum of the spacing of the major bedforms from simple to compound and complex types (Figure 6.23). This suggests that, given sufficient sand supply and time, simple dunes may grow into compound and complex varieties.

A GENERAL MODEL FOR DUNE SIZE AND SPACING

It appears therefore that there are no generally applicable relations between dune size or spacing and wind regime characteristics. The relations between dune size and potential sand transport rates indicate that the size of superimposed dunes increases with transport rates, whereas complex dune size decreases with transport rates. Therefore, it appears that the factors that control dune size and spacing are determined in part by the character of the dunes themselves. First, this is a result of the nature of dune dynamics. As the dune grows, it projects into the boundary layer and creates secondary patterns of air flow on and around itself. As discussed in Chapter 5, these secondary flows play a major role in the development of dunes by their control of patterns of erosion and deposition and the dynamics of super-imposed dunes. Second, there is a hierarchical system of aeolian dunes which consists of (1) individual simple dunes or superimposed dunes on compound and complex forms; and (2) compound or complex dunes (draas). There is a characteristic time period, termed the relaxation or reconstitution time (Allen 1974), over which each element of the hierarchy will adjust to changed conditions. Because change in dunes involves movement of sediment, an increasing spatial scale is therefore involved at each level of the hierarchy.

The reconstitution time can be represented by the time taken for the dune to migrate one wavelength in the direction of net transport. In the Namib Sand Sea, typical complex linear dunes have a spacing of 2,100 m and migrate at a rate of 0.05 m/yr. The reconstitution time for these dunes is therefore 42 ka. Crescentic dunes superimposed on the flanks of the linear dunes have a mean spacing of 90 m, and migrate at a rate of 3 m/yr, giving a reconstitution time of 30 yr. Reconstitution time therefore increases by several orders of magnitude from simple to complex dunes. This implies that the morphology of simple dunes and superimposed dunes is governed principally by annual or seasonal patterns of wind speed and direction and by spatial changes in wind speed over draas. The life span of these dunes is about 10 to 100 yr. Compound and complex dunes are relatively insensitive to seasonal changes in local air flow conditions and may persist for 1 to 100 ka. Their size is not a direct function of grain size or sand transport rates, but is the result of long-continued growth in conditions of abundant sand

Table 6.1 Orientation of dunes with respect to resultant and gross bedform-normal transport directions.

Dune type	Location	Dune orientation (°)	Maximum gross bedform-normal dune orientation (°)	Resultant transport direction (°)
Barchan	Namib	130	120	16
	Salton Sea	170	169	83
Crescentic (transverse)	Namib	130	120	16
		110	104	7
	Algodones	60	66	153
Linear				
Complex	Namib	19	118	42
		173	104	14
		3	112	24
		2	109	21
Simple	Sinai	110	119	102
	Kalahari	90	106	117
Star	Namib	176	174	16
	Gran Desierto	55	60	92

supply. The distribution of compound and complex dunes in sand seas is apparently controlled by regional-scale patterns of winds and sand transport rates. The Algodones dunes, California, are an excellent example of these principles. The compound crescentic dunes are approximately transverse to the mean annual resultant sand transport direction, whereas the superimposed dunes change their orientation in response to seasonal changes in wind direction (Havholm and Kocurek 1988).

The differences in scale between the elements of the aeolian bedform hierarchy suggest that the factors that control the size and spacing of simple dunes should be considered separately from those that influence the size and spacing of compound and complex varieties. Variations in winds and sand transport rates at different temporal and spatial scales appear to be the most important control of aeolian dune size and spacing. Whereas the height and spacing of individual simple and superimposed dunes in active sand seas probably tend toward an equilibrium with respect to contemporary sand transport rates and directions, the size and spacing of compound and complex dunes are functions of the long-term pattern of accumulation of sand in certain areas of the sand sea, determined by regional-scale patterns of winds and sand transport rates. As will be seen in Chapters 7 and 8, Quaternary climatic and sea level changes have had major effects on these patterns, and thus probably on the development of large complex dunes.

CONTROLS ON DUNE ALIGNMENT

The controls of dune orientation or alignment have long been a subject for speculation. Many workers (e.g. Aufrère 1928; Clos-Arceduc 1967) have concluded that dunes are oriented relative to the resultant or vector sum of sand transport. Thus dunes can be classified as transverse (strike of crestline approximately normal to resultant), longitudinal (crest parallel to resultant), or oblique (15–75° to resultant direction) (Mainguet and Callot 1978; Hunter et al. 1983).

Recently, Rubin and Hunter (1987) and Rubin and Ikeda (1990) have proposed on the basis of field and laboratory experiments with wind ripples and sub-aqueous dunes that all types of bedforms are oriented so that they maximize transport across the crest. This is the maximum gross bedform-normal rule. In this approach, transport from all directions contributes to bedform genesis and growth. By comparison, use of the resultant direction of transport as a control of bedform orientation implies that transport vectors from opposing directions cancel out each other and thus do not contribute to bedform growth (Figure 6.24).

Bedform trends in complex natural flows can be predicted by resolving each transport vector into bedform-normal and bedform-parallel components and determining the orientation that maximizes the gross bedform-normal transport. Using a sample of dunes from sand seas in Namibia and

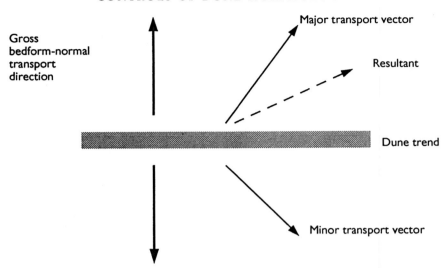

Figure 6.24 The concept of gross bedform-normal sediment transport (after Lancaster 1991).

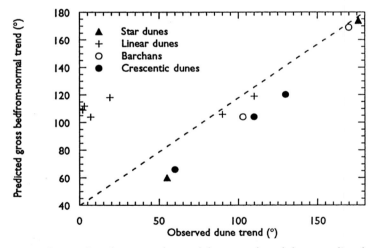

Figure 6.25 Comparison between observed dune trends and those predicted using the gross bedform-normal rule (after Lancaster 1991).

the southwestern United States (Table 6.1), Lancaster (1991) determined that there was a close agreement between observed and predicted bedform orientations in the case of many barchans, crescentic dunes, simple linear dunes, and star dunes (Figure 6.25), suggesting that all major dune types are oriented to maximize gross bedform-normal sediment transport. and therefore are dynamically similar. Differences between the form and orientation of dunes therefore result from variations in the directional characteristics of the flow, especially the angle(s) between the major transport directions and the ratio(s) between the magnitude of flows from different directions. The exception in this study is the complex linear dunes in the Namib Desert, which may indicate that the overall trends of these dunes were determined in some past wind regime, and that although they can be maintained in the present wind regime they are not in equilibrium with it, as has previously been suggested (Livingstone 1988). Given the very long reconstitution times of these large dunes, it is quite possible that they are not in equilibrium with current wind directions.

CONCLUSIONS

The modern process studies discussed in Chapter 5 provide an understanding of the relations between dune morphological types and environments revealed by studies of dune patterns on satellite images and aerial photographs. Sediment characteristics play a minor role and it is clear that the primary control of dune type is the local wind regime. Crescentic dunes occur in unimodal wind regimes, linear dunes in areas of bimodal winds, and star dunes in complex wind regimes, with dunes orienting themselves to maximize sediment transport normal to the crest. Vegetation cover is significant only in the formation of parabolic dunes, nebkha, and possibly some varieties of linear dunes.

There are good correlations between dune size and spacing that suggest some overall self-organization of dune patterns, but the nature of these controls are not well understood. There appear to be no generally applicable relations between dune size, spacing, and wind regime characteristics. Differences in scale suggest that the factors that control the size and spacing of simple dunes should be considered separately from those that influence the size and spacing of compound and complex varieties. Whereas the height and spacing of individual simple and superimposed dunes in active sand seas probably tend towards an equilibrium with respect to contemporary sand transport rates and directions, the size and spacing of compound and complex dunes are functions of the long-term pattern of accumulation of sand which is determined by regional-scale patterns of winds and sand transport rates that will be discussed in the next chapter.

7

SAND SEAS

INTRODUCTION

Sand seas are dynamic sedimentary bodies that form part of local- and regional-scale sand transport systems (see Plate 4) in which sand is moved by the wind from source zones to depositional sinks. They contain considerable volumes (km^3) of aeolian sand, together with local but significant amounts of interdune and extradune fluvial, lacustrine, and marine sediments. Many modern sand seas appear to have accumulated episodically in response to Quaternary climatic and sea level changes and probably represent an amalgamation of different generations of dune and interdune deposits (e.g. Kocurek *et al.* 1991; Lancaster 1992).

The global distribution of sand seas (Figure 7.1) indicates that they have accumulated in two main areas: (1) the tectonically stable desert regions of the Sahara, Arabian Peninsula, Australia, and southern Africa, and (2) the enclosed basins of central Asia. In these regions, sand seas cover between 15 per cent and 30 per cent of the area classified as arid. Much smaller amounts of sand are formed into the many small dune fields and sand seas of the Americas, where aeolian sand covers less than 1 per cent of the arid zone. In addition to modern active sand seas, extensive areas of relict vegetation-stabilized sand sheets and dune systems indicate that many sub-tropical sand seas were more extensive in the past, most notably during the last Glacial period (Bowler 1978; Sarnthein 1978; Lancaster 1990).

SAND SEA ACCUMULATION

A number of workers (e.g. Blandford 1876; Aufrère 1928; Capot-Rey 1947; King 1960; Capot-Rey 1970; Besler 1980) have noted the apparently close correspondence between the location of sand seas and low-lying areas in desert regions and argued they were formed by the reworking of underlying fluvial and lacustrine deposits. However, compilations of wind data together with information from aerial photographs and satellite images have shown that sand seas are the depositional sinks for regional and local scale sand

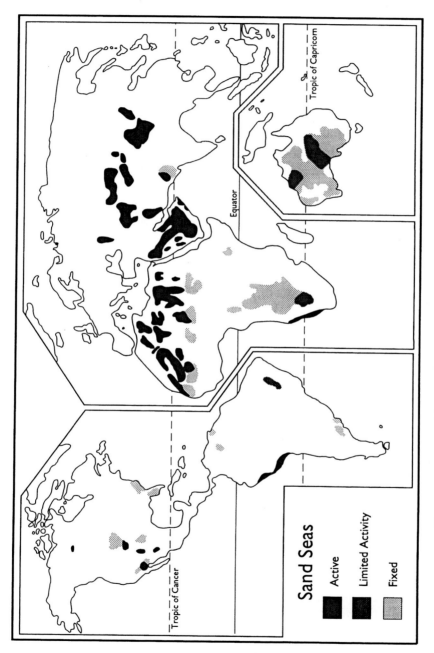

Figure 7.1 Location of the major sand seas of the world (after Thomas 1989).

transport systems (Wilson 1971; Fryberger and Ahlbrandt 1979). Long-distance transport of sand by the wind occurs in the Sahara, Arabian Peninsula, Mojave, and Namib deserts (e.g. Lancaster and Ollier 1983; Mainguet and Chemin 1983; Fryberger *et al.* 1984; Mainguet 1984b; Zimbelman and Williams 1990), whereas Australian sand seas and many North American dune fields receive sand from local sources (Norris and Norris 1961; Sharp 1966; Wasson *et al.* 1988; Blount and Lancaster 1990).

Models for the development of sand seas

As with dunes, concepts of sediment budgets can be applied to sand seas to understand their accumulation. Accumulation of sand is the product of spatial changes in transport rates and temporal changes in sediment concentration such that:

$$dh/dt = -(dq_s/dx + dC/dt) \qquad (7.1)$$

where h is the elevation of the deposition surface, t is time, q_s is the spatially averaged bulk volume sediment transport rate, x is the distance along the transport pathway, and C is the concentration of sediment in transport. In this model, the transport rate (q_s) consists of two components: that due to bedform migration (bedform transport, q_b) and that which is throughgoing (q_t) as a result of saltation or bedload transport. The sediment concentration (C) is a measure of the total amount of sediment in transport, and can be approximated by the average height of the dunes (Kocurek and Havholm 1994).

Although sediment transport rates over sand surfaces are almost always at the capacity of the wind to transport sediment (sediment saturated), those over other desert surfaces (exposed bedrock, alluvial fans, playas) are frequently below transport capacity (undersaturated or metasaturated in the terms of Wilson (1971)), because sediment supply is limited. Thus actual sand transport rates (q_a) may be less than potential rates (q_p) in proportion to the ratio q_a/q_p, which ranges between zero for completely undersaturated flows to 1 for fully saturated conditions. The wind is thus potentially erosional until its transport capacity is reached, regardless of whether the wind is accelerating, steady or decelerating. Deposition occurs wherever there are local decreases in transport capacity (e.g. deceleration in the lee of obstacles or changes in surface roughness). The wind may still be transporting sand, as deposition only occurs until the transport rate is in equilibrium with changed conditions.

Following principles of sediment mass conservation, if transport rates decrease in the direction of flow, deposition will occur and the accumulation will grow upwards. If, however, sediment transport rates increase, then the accumulation will be eroded. No change in transport rates in space will give rise to net sediment bypassing (Kocurek and Havholm 1994). The relations

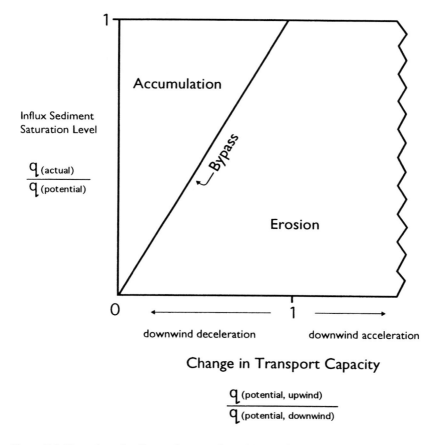

Figure 7.2 Domains of sediment bypass, deposition and erosion (after Kocurek and Havholm 1994).

between the sediment saturation level of the input to the system and the change in transport capacity in time and space (defined as the ratio between the potential transport rate upwind and downwind of the area of concern) determine the domains of erosion, bypass, and accumulation of sand (Figure 7.2). These concepts can be used to assess the nature of changes that may occur through time. For example, assuming that the basic wind regime characteristics of an area do not vary with time, changes in sediment availability from source zones and thus the saturation level of the input will determine the behaviour of the system.

Changes in sediment budgets through time may occur as a result of climatic changes, eustatic changes in sea level, or tectonic causes that affect

201

sediment supply and/or mobility, and may lead to the episodic development of sand seas. One manifestation of this type of change is the development of super- or regional-scale bounding surfaces (Kocurek 1988; Kocurek and Havholm 1994) that truncate aeolian accumulations and therefore represent periods when accumulation ceased, to be replaced by sediment bypassing, geomorphic stability or deflation over part or all of the area of a sand sea.

The development of sand seas is ultimately limited by sand availability from source areas and long term sediment budgets. Not all sand seas develop to the point where accumulation of sand (defined as net deposition over time) takes place. There are numerous instances (e.g. parts of the southwest Kalahari, Simpson-Strzelecki, Akchar Erg, Skeleton Coast, and Algodones dunefields) of dunes that directly overlie bedrock and/or alluvial deposits. These represent sand seas that have not yet generated an aeolian accumulation. By contrast, the Namib Sand Sea, the Occidental and Oriental Ergs, and the Issaouane-N-Irraren of the Sahara (Wilson 1973) as well as Kelso Dunes and the Great Sand Dunes have accumulated at least 20–30 m of sand below the present dune forms.

Wilson (1971) recognized four main factors that influence the growth of sand seas with bedforms: (1) deposition takes place even when sand flow is below transport capacity; (2) sand sea growth is initially by lateral extension rather than vertical accretion; (3) sand sea shape and actual sand flow rates and deposition patterns are time dependent until bedform growth is complete; and (4) bedform type influences sand sea development: sand seas which are dominated by 'sand passing' dunes such as linear dunes develop by bedform extension at their downwind margins and grow slowly, whereas sand seas with a high proportion of 'sand trapping' bedforms (crescentic and star dunes) grow by both bedform migration at the downwind margin and bedform growth at the upwind margin. Most sand seas are a combination of 'sand passing' and 'sand trapping' bedforms.

The basic mechanism by which all sand seas accumulate is bedform climbing (Rubin and Hunter 1982), which involves upward movement relative to the generalized depositional surface (a line midway between the crests and troughs of the dunes). Bedform climbing occurs principally as a result of decreases in sediment transport rates in the direction of dune migration (Rubin and Hunter 1982). In the case of migrating dunes leaving a net deposit, each successive dune must climb over the accumulations of the preceding dune (Figure 7.3).

Following Rubin and Hunter (1982), climb in the simple case of migrating two-dimensional asymmetric dunes occurs because sediment deposited on the lee slopes is not completely eroded on the stoss slope as migration takes place (Figure 7.3). The nature of airflow over dunes and interdune areas (see Chapter 5) indicates that the lower stoss slopes of dunes and interdune areas are potentially erosional until dune growth has closed the interdunes (Kocurek and Havholm 1994), as first suggested by Wilson (1971). This state

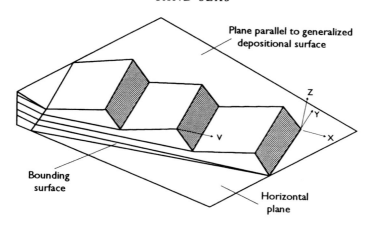

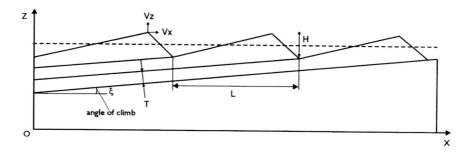

Figure 7.3 Bedform climbing (after Rubin and Hunter 1982). H is bedform height, L is bedform spacing, T is thickness of translatent stratum, V is direction of bedform migration with components Vx and Vz.

must be reached before bedform climbing and aeolian accumulation can take place and can be observed in areas of crescentic dunes at Great Sand Dunes (Andrews 1981), Kelso Dunes (Lancaster 1993b), the coastal parts of the Namib Sand Sea (Lancaster 1989b), and in the downwind areas of the Skeleton Coast dune field (Lancaster 1982a). The equivalent in areas of linear and star dunes may be the sand-covered interdunes that dominate in central areas of the Namib Sand Sea.

Relations between sand seas and regional sediment transport systems

Concepts of sediment mass conservation indicate that deposition of sand and the accumulation of sand seas will occur downwind of source zones wherever sand transport rates are reduced as a result of changes in climate or topography. Sand seas are therefore located in areas where sand transport rates are lower and/or more variable in direction, compared to adjacent areas without sand sea development (Fryberger and Ahlbrandt 1979). Wilson (1971) recognized three situations in which sand seas might develop: (1) where sand transport from all directions converges (e.g. areas of multidirectional winds); (2) where sand transport from two opposite directions converges (e.g. in seasonally reversing wind regimes); and (3) where transport pathways cross the sand sea, with local convergence and deceleration of the wind giving rise to deposition.

In the Sahara and Arabian Peninsula, sand is moved long distances between sand seas on pathways that are characterized by sand-choked river valleys, shadow dunes and barchans, small climbing and falling dunes, and sand streaks and sheets, and are clearly visible on satellite images by their bright tones (Breed *et al.* 1979; Mainguet 1984b). In the Sahara, the sand transport

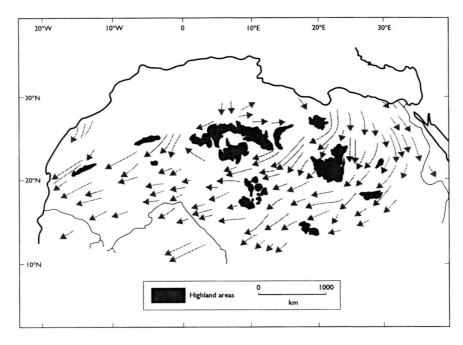

Figure 7.4 Sand seas and sand transport pathways in the Sahara (after Mainguet 1984b).

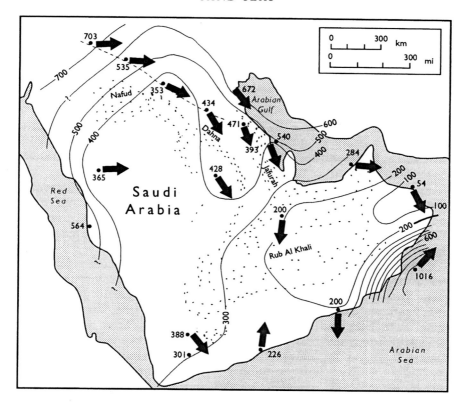

Figure 7.5 Sand transport pathways and wind systems in the Arabian Peninsula. Isolines denote RDP (after Fryberger *et al.* 1984).

pathways (Figure 7.4) reflect the dominance of the trade wind circulation that tends to move sand from the sand-poor eastern and central parts of the desert towards thick sand accumulations in the Sahel, and from the piedmont of the Atlas Mountains and the Mediterranean coast to the northern sand seas (Mainguet and Chemin 1983). Sand transport pathways in the Arabian Peninsula (Figure 7.5) reflect the importance of the west to northwest Shamal winds that transport sand from Iraq and the Gulf coast towards the Rub' al Khali (Fryberger *et al.* 1984). The Wahiba sand sea forms part of a separate system dominated by the southwest monsoon (Warren 1988). In the Namib, sand is moved inland from coastal source areas, fed ultimately by the Orange River, to accumulate in large complex linear and star dunes in the centre of the sand sea (Lancaster 1989b). Regional-scale aeolian transport corridors in the Mojave Desert transfer sand from western and northern parts of the region towards sediment sinks in the eastern Mojave (Smith 1984; Zimbelman and Williams 1990). However, not all sand seas accumulate in this manner. Regional-scale patterns of dune size, sediment thickness, and composition in Australia show clearly that local sources are dominant, although there is a continent-wide linear dune system (Wasson *et al.* 1988).

Effects of topography

Some sand seas accumulate where sand transport pathways converge in the lee of topographic obstacles. In the southern Sahara, the Fachi-Bilma Erg has accumulated where the northeast trades converge in the lee of the Tibesti Mountains (Mainguet and Callot 1978). The Akchar Erg of Mauritania lies downwind of the Adrar Escarpment (Kocurek *et al.* 1991), and the Skeleton Coast dune field of Namibia is anchored in the lee of low hills (Lancaster 1982a). Areas of low elevation also favour sand sea development by enhancing convergence in the regional wind field (Wilson 1971) and the Mallee and Simpson-Strzelecki dune fields of Australia lie in topographic basins (Wasson *et al.* 1988).

Other sand seas occur on the upwind sides of topographic obstacles where winds are checked by topographic barriers. Observations and models of wind flow over escarpments (e.g. Bowen and Lindley 1977; Pearse *et al.* 1981) indicate that wind velocity is reduced at their base. The location of many dune fields and sand seas adjacent to mountain fronts was probably originally determined by wind velocity and transport rate reductions of this nature, leading to deposition of sand upwind of the mountain block. Once established, feedback between the growing sand accumulation and the wind moves the point of minimum velocity upwind, so that the dune field progrades in this direction (Mainguet 1984b). Parts of the Grand Erg Oriental in the Sahara and Great Sand Dunes, Colorado, are examples of the operation of such a mechanism (Fryberger and Ahlbrandt 1979). Likewise, many dune fields in the Basin and Range Province of the United States (e.g. Kelso, Eureka dunes) owe their location to this type of process, although the effects of opposed winds coming through mountain passes may also play a role (e.g. Sharp 1966).

Effects of regional changes in winds

Reduced sand transport rates may be the result of changes in regional circulation patterns that decrease wind speeds and/or increase their directional variability. Some sand seas, such as the Gran Desierto of Mexico (Figure 7.6) and the northern Saharan sand seas, occur in areas in which wind directions change seasonally, resulting in input of sand from two directions, and growth of star and reversing dunes (Breed *et al.* 1979; Mainguet and Chemin 1983; Lancaster *et al.* 1987). Elsewhere, as in the Jafurah of Saudi Arabia (Fryberger *et al.* 1984) and the Akchar of Mauritania, sand seas are crossed by sand transport pathways and deposition takes place as a result of

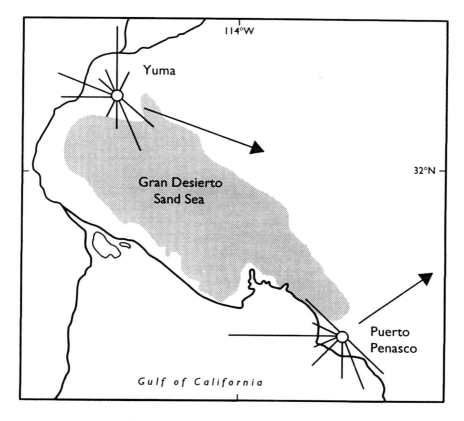

Figure 7.6 Patterns of winds and potential sediment transport in the Gran Desierto Sand Sea.

local flow convergence or regional decreases in wind energy.

The accumulation of the Namib Sand Sea is a product of both regional decreases in wind energy and a parallel increase in directional variability. Sand is transported from southern and western coastal areas which are areas of low directional variability and high wind energy towards central, northern and eastern parts of the sand sea, where winds are more variable in direction (Figure 7.7). As sand transport rates decrease in the mean transport direction, deposition occurs. Sand is also being transported into areas where winds are opposed in direction. Consequently, once sand reaches such areas, it will tend to remain there and add to the dunes. These are the areas of large dunes and thick sand cover.

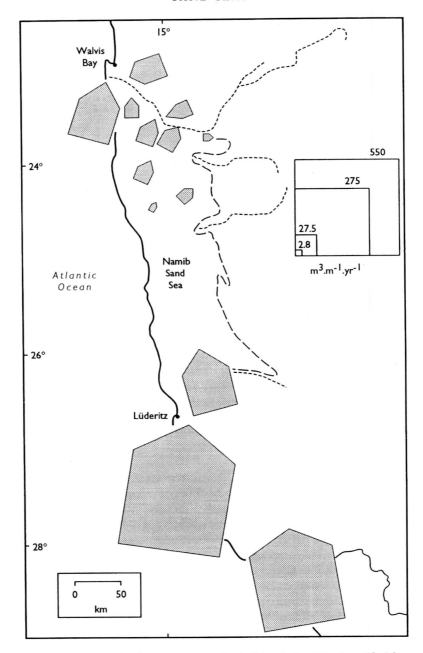

Figure 7.7 Patterns of sediment transport in the Namib Sand Sea (modified from Lancaster 1985a).

DUNE PATTERNS IN SAND SEAS

The pattern of dunes of different types and the spatial variation in their size, spacing and alignment in a sand sea are the surface expression of the factors that control its dynamics and accumulation. Many sand seas show a clear spatial patterning of dune types, dune size and spacing, and sediment thickness (Breed *et al.* 1979; Porter 1986) illustrated schematically in Figure 7.8. Upwind near-source areas are characterized by sand sheets and zibars, and areas of low crescentic dunes. Sediment thickness is low. The central areas of the sand sea are occupied by large compound and complex dunes that represent the major area of sediment accumulation. Downwind, the leading

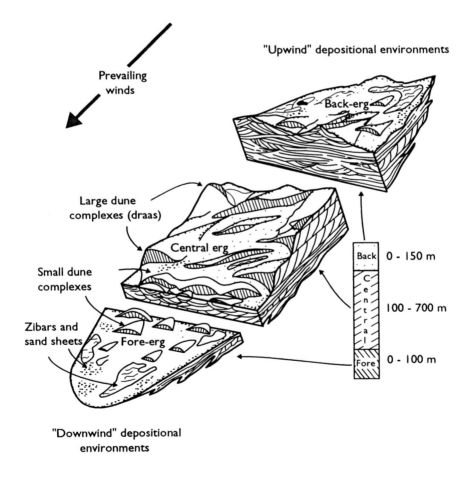

Figure 7.8 Schematic illustration of sand sea deposits (after Porter 1986).

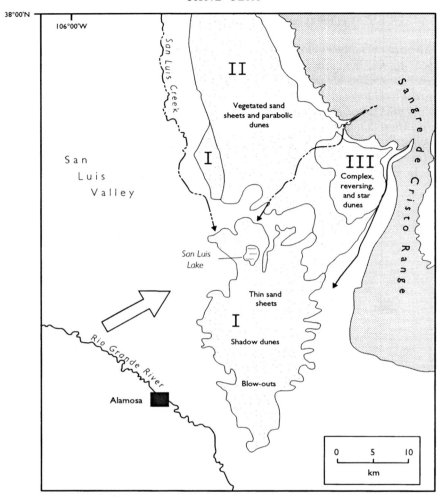

Figure 7.9 Sedimentary provinces at Great Sand Dunes, Colorado (after Andrews 1981).

edge of the sand sea is an area of thin sediment accumulations with small, mainly crescentic dunes and prograding sand sheets and streaks.

At Great Sand Dunes, Colorado, Andrews (1981) recognized three dune provinces (Figure 7.9) that lie along the direction of sand transport. Province I is characterized by a thin sand cover (less than 1 m), a high water table and a series of ephemerally flooded and seasonally flooded alkaline lakes. Aeolian deposits in this area are partly or completely cemented by sodium bicarbonate, calcite, and other evaporite minerals, and include 5–10 m high 'berms' or

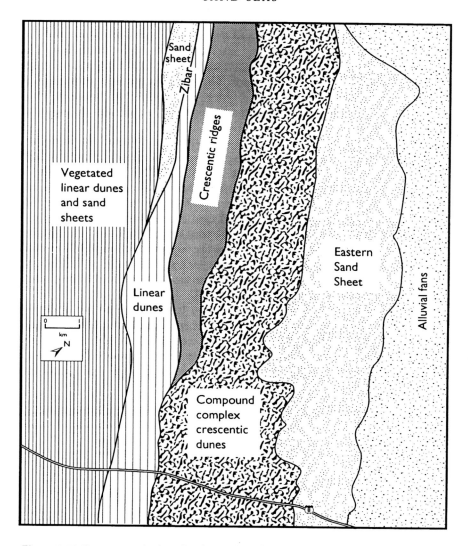

Figure 7.10 Dune types in the Algodones dune field (redrawn from aerial photograph in Havholm and Kocurek 1988).

vegetated lunette dunes on the eastern margins of the lake basins, active sand sheets, shadow dunes, and blow-outs. Province II consists of vegetation-stabilized sand sheets, parabolic dunes, blow-outs, and shadow dunes, with small areas of active crescentic dunes and granule-rippled sand sheets, with a total thickness of as much as 10 m. Province III is a complex accumulation of dunes that is as much as 100–180 m thick and consists of 5–15 m high crescentic dunes along the margins of the dune mass, 30–70 m high reversing crescentic dunes in the central parts of the dune field, and small areas of star dunes on the southern, eastern, and northern edges of the dune field. The dunes are superposed on the sand accumulation so that the highest crests are 200 m or more above the valley floor. The complex patterns developed in Province III result from the coalescence or shingling of dunes as their migration towards the east is restricted by topographic barriers and opposed wind directions.

The Algodones Dunefield of southeast California (Figure 7.10) is characterized by a core of compound to complex crescentic ridges up to 60 m high separated by deflated interdunes as much as 500 m wide. Forming a transitional zone on the east and west margins of the compound dunes is a zone of simple crescentic dunes. The east side of the dune field consists of small crescentic ridges and barchans and sand sheets that is prograding onto the adjacent alluvial fans. By contrast, the west side of the dune field consists of a series of simple linear dunes that lie at the top of a 35 m high, 500 m wide ramp of coarse sands and zibars that represents an accumulation left behind as the dune field has migrated slowly to the east (Norris and Norris 1961; Nielson and Kocurek 1986; Sweet *et al.* 1988).

Dune patterns in Saharan sand seas have been described by Mainguet (1972, 1984a, b); Mainguet and Callot (1978); and Breed *et al.* (1979) and are summarized in Table 7.1. The northern Saharan sand seas (Figure 7.11) are characterized by complex crescentic and star dunes developed in a multidirectional wind regime that results from the interaction of trade wind and

Table 7.1 Percentage of area covered by major dune types in Saharan sand seas.

Dune type	North	South	Northeast	West
Crescentic				
Simple and compound	33.4	28.4	14.5	19.2
Complex	26.1	24.3	12.6	18.5
Linear				
Simple and compound	5.7	24.1	2.4	35.5
Complex	13.5		13.5	
Star	7.9		23.9	
Sand sheets and streaks	36.9	47.5	39.3	45.3
Other/undifferentiated		5.3		

Sources: Data from Fryberger and Goudie 1981, with additions from Lancaster 1989b.
Note: The original data from Landsat MSS images had a spatial resolution of 80 m per pixel. As a result, many dunes classified as simple may be compound forms and some of the sand sheet areas probably represent areas of low dunes.

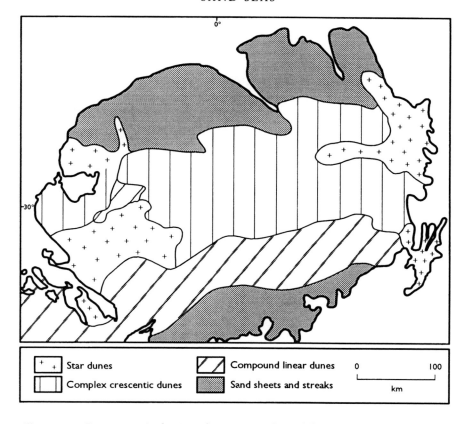

Figure 7.11 Dune types in the Grand Erg Oriental (modified from Breed *et al.* 1979).

Legend:
- Star dunes
- Complex crescentic dunes
- Compound linear dunes
- Sand sheets and streaks

0   100
km

mid-latitude circulations. By contrast, sand seas in the western and southern Sahara are mainly composed of linear dunes, with small areas of crescentic dunes. The dominance of linear dunes in these areas reflects the persistence of trade wind circulations. Aeolian accumulations in the eastern Sahara include linear and crescentic dunes (Haynes 1989), but are dominated by extensive sand sheets and areas of gently undulating dunes with a chevron pattern (Maxwell and Haynes 1989). The formation of extensive sand sheets is attributed to the abundance of coarse sand derived from the Mesozoic Nubian Sandstone that underlies much of these areas (Breed *et al.* 1987).

Continent-wide patterns of dunes in Australian sand seas and dune field have been documented by Wasson *et al.* (1988). Although linear dunes dominate, short, narrow-crested varieties appear to occur where sand transport is limited by vegetation cover; broad-crested linears are found in the northern parts of the dune system, and narrow-crested varieties in the south. Network dunes occur in small basins in the central parts of the

213

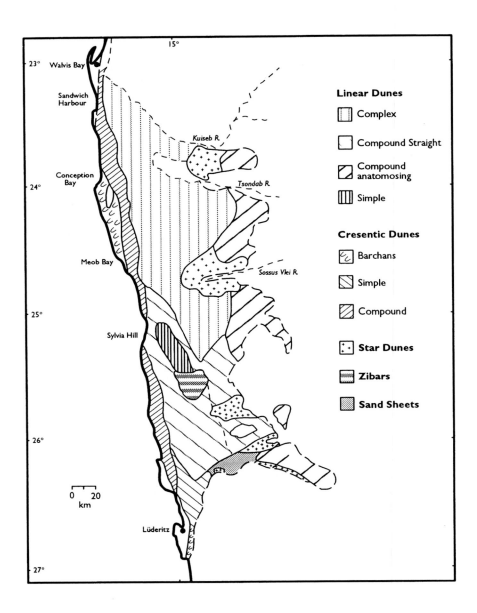

Figure 7.12 Patterns of dune morphology in the Namib Sand Sea (after Lancaster 1989b).

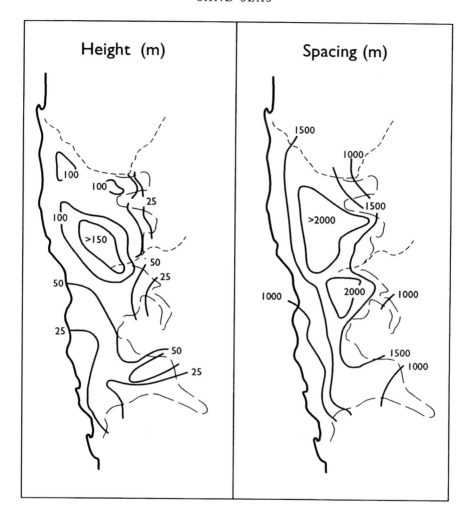

Figure 7.13 Variation of dune height and spacing in the Namib Sand Sea (after Lancaster 1989b).

continent where the directional variability of the wind regime is high. Wasson *et al.* (1988) found that the highest and most widely spaced dunes occur on coarse substrates and have the thickest equivalent sand cover. Mapping of patterns of sediment thickness, together with sedimentological and mineralogical evidence showed that most sand in the Australian deserts has only moved short distances from its source, although the dune pattern suggests an organized continent-wide sediment transport system.

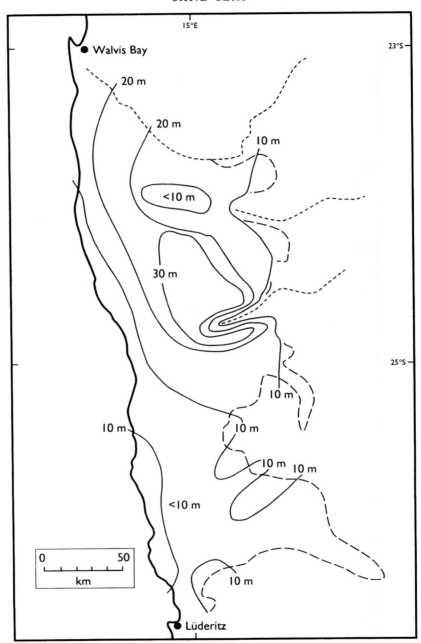

Figure 7.14 Spatial variation of sediment thickness in the Namib Sand Sea (after Lancaster 1989b).

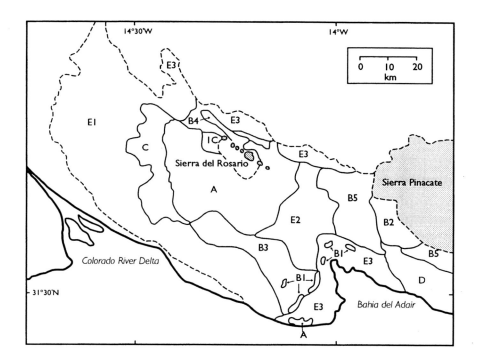

Figure 7.15 Patterns of dune types in the Gran Desierto Sand Sea.
    A: Chains and clusters of star dunes.
    B: Crescentic dunes:
       1, 2 Active simple crescentic dunes.
       3 Large relict crescentic dunes, stabilized by vegetation.
       4 Coalescing multiple generations of largely relict crescentic dunes.
       5 Compound/complex crescentic dunes.
    C: Reversing dunes with active crescentic dunes on margins of the area,
       with stabilized and relict crescentic dunes in topographically lower
    areas.
    D: Linear and parabolic dunes, largely vegetated.
    E: Sand sheets:
       1 Sparsely vegetated, low relief.
       2 Moderately vegetated, 2–3 m local relief.
       3 Undifferentiated.

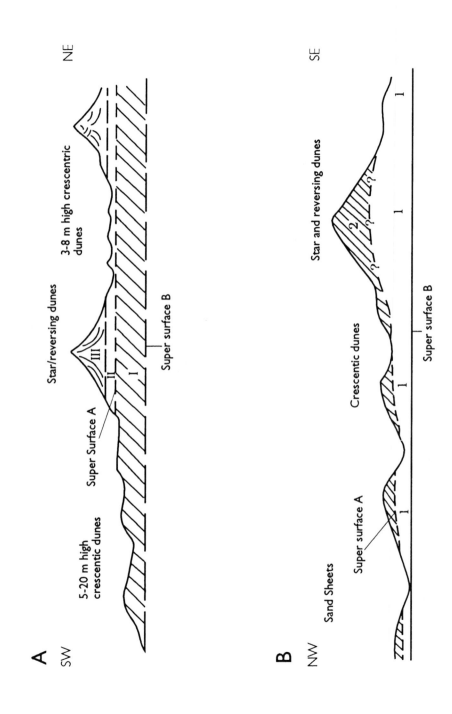

A

SW   NE

Star/reversing dunes

3-8 m high crescentric dunes

5-20 m high crescentic dunes

Super Surface A

Super surface B

III

II

I

B

NW   SE

Sand Sheets

Star and reversing dunes

Crescentic dunes

Super surface A

Super surface B

1

1

1

2

?

?

?

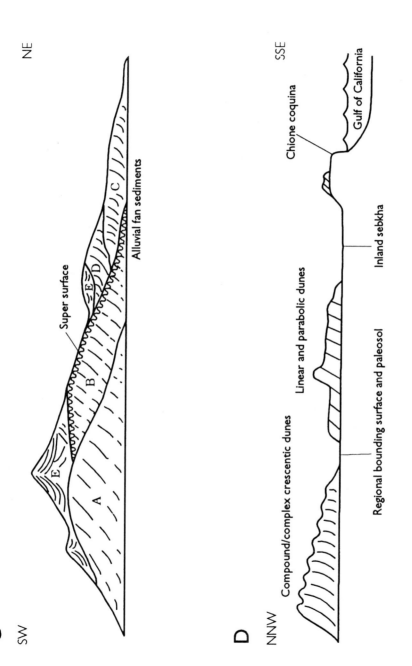

C

SW

NE

E

B

A

E

D

C

Super surface

Alluvial fan sediments

D

NNW

SSE

Compound/complex crescentic dunes

Linear and parabolic dunes

Chione coquina

Gulf of California

Regional bounding surface and paleosol

Inland sebkha

Figure 7.16 Dune generations in the Gran Desierto Sand Sea.

The Namib and Gran Desierto sand seas provide contrasting case studies of dune patterns in sand seas developed in multidirectional wind regimes. The Namib Sand Sea (Figure 7.12) is characterized by a well-organized dune pattern in which crescentic dunes occur in unidirectional wind regimes near the coast; compound and complex linear dunes are associated with bi-directional winds inland; and star dunes with multidirectional wind regimes along the eastern margin of the sand sea (Lancaster 1983a). Dune height and spacing vary together in a systematic way in the Namib Sand Sea (Figure 7.13). Dunes are highest and most widely spaced in the central and some northern parts of the sand sea, with progressively lower and more closely spaced dunes towards the margins. The thickness of sand over most of the central parts of the sand sea exceeds 20 m and is more than 30 m north and northwest of Sossus Vlei (Figure 7.14). Apart from small areas of star dunes, equivalent sand thickness in the southern parts of the sand sea is less than 10 m. Patterns of dune morphology, sand thickness, and sediment characteristics (see below) suggest that the Namib Sand Sea represents a single integrated depositional system.

By contrast to the Namib Sand Sea, a very characteristic feature of the Gran Desierto Sand Sea (Figure 7.15) is the juxtaposition of dunes of contrasting composition and morphology (Lancaster et al. 1987; Lancaster 1992). In many areas, active star dunes occur next to stabilized crescentic dunes. In other areas of the sand sea several sets of crescentic dunes, each with a different morphology and alignment, coalesce with each other (Figure 7.16). Analysis of the pattern of dunes reveals that there are two scales of variability in dune morphology in this sand sea: (1) at the regional scale, there is a progression from sand sheets in the northwest through a zone of crescentic and reversing dunes to the central star dune area. The eastern part of the sand sea is characterized by large compound/complex crescentic dunes, as well as areas of vegetated sand sheets and linear and parabolic dunes; and (2) at the local scale, different dune generations are juxtaposed or superposed on each other. Spatial changes in dune morphology at the first scale are characterized by transitions from one dune type to another (e.g. from crescentic through reversing to star dunes) as described by Lancaster (1989a). They are the result of regional changes in wind regimes (see Figure 7.6) from one dominated by northerly and westerly winds to the west and north of the sand sea (sand sheets and crescentic dunes migrating towards the southeast) and southerly winds to the east and south (crescentic dunes migrating to the north and northwest). The star dunes occur in the zone where these two wind regimes interact with each other. Other changes are partly due to the effect of increasing dune size on bedform reconstitution time in a seasonally varying wind regime: small dunes can be reformed in a single wind season, whereas larger dunes exhibit a morphology that is controlled by several wind directions. The extensive sand sheets probably represent a response to limited sediment supply and sediment bypassing.

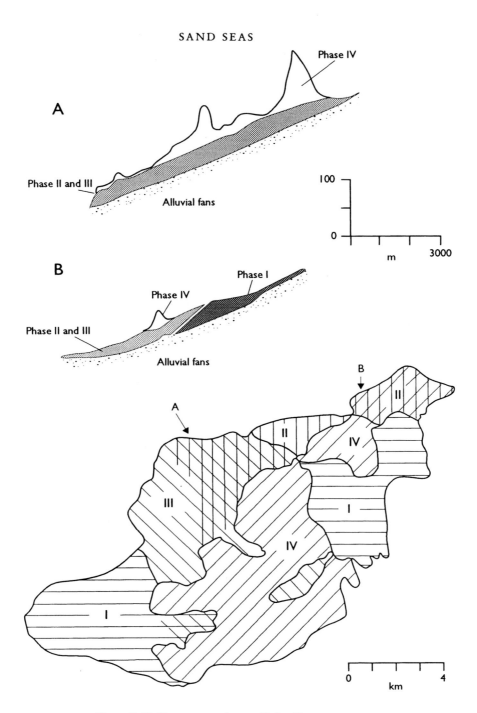

Figure 7.17 Dune generations at Kelso Dunes.

However, the juxtaposition of dunes with different morphologic types, alignments, grain-size composition, and degree of post-depositional modification of sedimentary structures can best be explained by regarding each dune morphologic unit as a separate and genetically distinct generation of dunes. The Gran Desierto sand sea therefore represents an amalgamation of multiple generations of dunes that were deposited adjacent to each other or superposed on one another.

Similar patterns of multiple dune generations occur in many other sand seas. In the Wahiba Sand Sea, there is an older pattern of 50–100 m high complex linear dunes inland, with low crescentic and network dunes near the coast (Warren 1988). Linear dunes in the Simpson-Strzelecki Desert are of two generations formed in different climatic and hydrological conditions: reddened quartz-rich linear dunes formed prior to 20 ka, and younger pale linear dunes (Wasson 1983; 1984). At Kelso dunes in the Mojave Desert, four main dune generations can be recognized on the basis of composition, morphology, and alignment patterns (Lancaster 1993b). Geomorphic relationships between the different dune morphological units suggest that Kelso Dunes represents in part a stacked sequence of dunes of different generations (Figure 7.17). 'Stacking' of dune generations has occurred because the expansion of the dune field in response to sediment inputs is restricted by its location on the piedmont of the Granite Mountains.

GRAIN SIZE AND SORTING PATTERNS IN SAND SEAS

In parallel with changes in dune types, there are distinct trends in the size and sorting character and composition of dune sediments in many sand seas (e.g. Sweet et al. 1988; Lancaster 1989b). Sands in upwind areas tend to be coarser and less well sorted than those in the depositional centres of many sand seas. Progressive fining of sands and improvements in sorting along the net transport direction have been noted from Great Sand Dunes (Wiegand 1977); the southwestern Kalahari (Lancaster 1986) (Figure 7.18), the Skeleton Coast dune field (Lancaster 1982a), the Wahiba Sands (Allison 1988) and the Algodones Dunes (Sweet et al. 1988). These patterns result from the progressive concentration of the slower moving, coarse reptation and creep populations in upwind, near source areas, which also explains the common occurrence of sand sheets and zibars in the upwind parts of sand seas.

In the Namib Sand Sea, there is a consistent pattern in grain size and sorting parameters (Figure 7.19) with fine, well sorted, near symmetrical sands in central and northern areas of the sand sea and coarser, less well sorted sands in southern and some western parts. There is also an area of coarser, but very well sorted sand, centred on Sossus Vlei. The patterns are best explained by sand movement away from source zones to the south and west of the sand sea during which coarse grains are left behind in upwind areas.

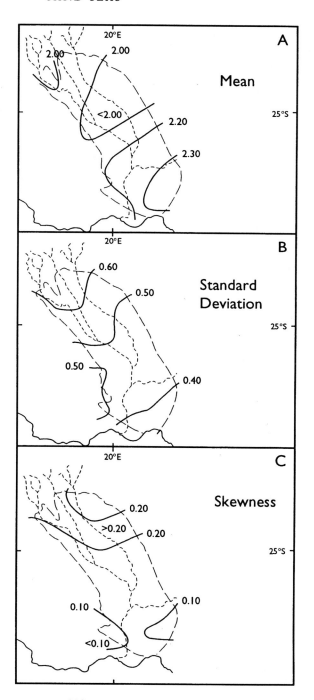

Figure 7.18 Spatial variation of grain size and sorting parameters in the southwestern Kalahari linear dune field (after Lancaster 1986).

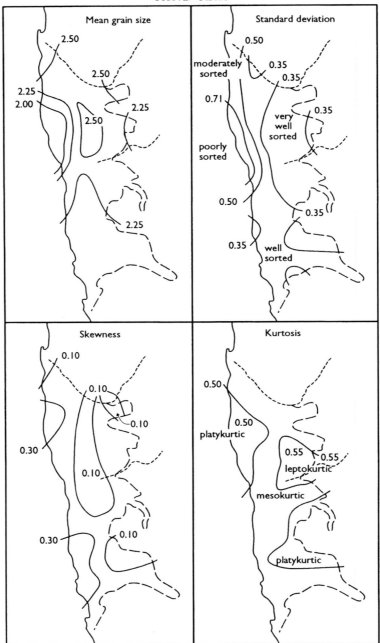

Figure 7.19 Spatial variation of grain size and sorting parameters in the Namib Sand Sea (after Lancaster 1989b).

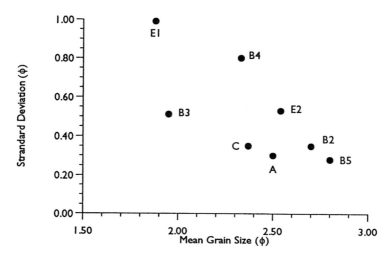

Figure 7.20 Comparison of grain size and sorting parameters of different dune generations in the Gran Desierto Sand Sea (for locations see Figure 7.15).

In the Gran Desierto, by contrast, each dune morphologic unit is associated with sands that are texturally and compositionally distinct (Figure 7.20) and there is no consistent regional-scale pattern of grain size and sorting (Blount and Lancaster 1990). The observed textural and compositional patterns appear to be best explained in terms of a series of episodes of sand supply from different source areas. In order of importance these were: (1) fluvial and deltaic sediments of the Colorado River, and (2) beaches of the Gulf of California, and alluvial fans and ephemeral streams. The eastern compound crescentic dunes are composed of very well sorted, very fine sand and were derived from a southern coastal source zone, whereas much of the sand in the star dune area is coarser and less well sorted and came from the Colorado River delta and its former floodplain. Similar patterns in which distinct sedimentary characteristics are associated with morphologically distinct dunes occur in a number of other sand seas including the Simpson-Strzelecki sand sea (Wasson 1983) and at Kelso Dunes (Lancaster 1993b).

MODES OF SAND SEA ACCUMULATION

Detailed studies of dune patterns and stratigraphy in sand seas provide a model for sand sea development in which the history of the sand sea determines its form. Of particular importance are the effects of climatic and sea level changes on sediment supply and dune formation (discussed in detail in Chapter 8). Sand seas that have a single main sediment source and have been affected to a limited degree by external changes tend to have well-organized dune patterns in which changes in dune morphology can be related to regional changes in wind regimes. A good example of this type of sand sea is the Namib Sand Sea. At the other extreme are the complex dune patterns of the Gran Desierto and the Wahiba Sands where multiple sediment sources and complex climatic and sea level changes have resulted in the development and preservation of many different dune generations. Although there is local erosion in interdune areas and on stoss slopes of dunes, preservation of older generations of dunes is likely because sand seas and dune fields are areas of net accumulation or at least sediment bypassing (otherwise they would not exist). Preservation of older dune generations is favoured by climatic changes that result in surface stabilization by vegetation and soil formation.

Within this framework, two end member models of sand sea accumulation can be recognized: (1) stacking of morphologic and genetic units (dunes and/ or sand sheets). This scenario leads to thickening of the evolving sand body as successive genetic units (which may be separated by regional scale or 'super' bounding surfaces) are superposed on one another; and (2) mosaicking of genetic units whereby successive morphologic and/or genetic units are deposited adjacent to each other. This leads to expansion of the area of the sand sea with little thickening of the accumulation, and poor preservation of any super surfaces unless they are stabilized by soil formation, surface lags, or cementation. This process will only occur if there is a significant time interval between the accumulation of each unit. Different dune generations, however, will be preserved largely intact.

The controls on the mode of sand sea accumulation appear to be sand supply and accumulation rate. Stacking will occur when sand supply is high; bedform climbing is rapid, implying a sharp decrease in transport rates in the direction of transport; and/or the area of the accumulation is restricted by topography, as at Kelso Dunes, California (Lancaster 1993b) and Great Sand Dunes, Colorado (Andrews 1981). Sand seas that accumulate as a result of converging sand transport pathways in multidirectional wind regimes may also develop in this way. Mosaicking will take place in situations of low sand supply and accumulation rates, where bedform migration and extension dominate over vertical accretion. This mode of sand sea accumulation appears to be most common in sand seas in the Sahara, Arabia, Australia, and southern Africa that are developed on tectonically stable cratons. Many of these sand seas have very thin deposits, and dunes rest on the underlying bedrock substrate (Kocurek et al. 1991).

CONCLUSIONS

Sand seas are the major accumulations of aeolian sand and form part of regional-scale sediment transport systems. Their formation can be explained by reference to principles of sediment mass conservation similar to those that determine dune formation and dynamics, but the time scales involved are much longer. Relations between the sediment saturation level of the incoming sediment stream and changes in transport rates in time and space determine the domains of erosion, deposition, and bypass of sand.

Sand seas are fed from source areas via local and regional scale sand transport corridors and accumulate in areas where regional wind regimes become more variable in direction or less energetic, or where transport pathways encounter topographic obstacles. Patterns of dunes in sand seas reflect the factors that have affected its accumulation. Upwind areas are characterized by thin sand cover and sediment bypassing or erosion, central areas by thicker sand cover and more complex dune types, and the leading edge by small migrating or extending dunes. Some sand seas have well-integrated dune patterns whereas others are characterized by the juxtaposition of dunes of different morphologic and/or sedimentary characteristics. Such patterns suggest episodic accumulation in response to external changes in sediment supply and mobility. As sand seas comprise large volumes of sand and have taken thousands to tens of thousands of years to accumulate, environmental changes during the Quaternary era have played a major role in sand sea evolution. The effects of these changes will be discussed in the following chapter.

8

PALAEOENVIRONMENTS AND DUNES

INTRODUCTION

Many sand seas and dune fields are composed of large volumes of sediment and have accumulated over periods of 10^3 to 10^5 yr, during which Quaternary climatic and sea level changes have had a major influence on sand supply and dune mobility. As a result, many sand seas and dune fields show evidence of episodic accumulation. Dunes are also important sources of palaeoclimatic data. Wind action is enhanced by dry conditions and extensive areas of bare soil, therefore the existence of relict aeolian deposits and landforms has long been regarded as an indicator of past aridity. Aeolian landforms and deposits have the potential to provide information on the geographical extent, duration, and timing of periods of enhanced aridity, as well as changes in wind directions and circulation patterns over time. Data on the latter are not available from any other source.

Because dunes are the result of interactions between the wind and the land surface, they are sensitive to changes in both atmospheric parameters and surface conditions. Aeolian action is therefore influenced by changes in wind strength and direction that may be the direct result of global or regional climatic changes as well as by changes in vegetation cover, soil moisture, and sediment availability that are an indirect effect of climatic change.

Sand seas can provide palaeoenvironmental information from three main sources: (1) relict dune systems; (2) interdune deposits; and (3) penetration of extra-dune fluvial deposits into the margins of the sand sea. The quality of palaeoclimatic data provided from sand seas is often much better than that obtained from adjacent rocky desert regions, as sand dune areas tend to accentuate the effects of both dry and wet phases (Rognon 1982), and respond rapidly to changes in climate (Figure 8.1). In periods of increased rainfall, the high infiltration capacity and porosity of dune sands favour the growth and persistence of vegetation, which may lead to the partial or complete stabilization of the dunes, and the formation of soils in dune sands. If the periods of increased rainfall are of sufficient magnitude and duration, then water tables will rise, leading to the accumulation of pond and marsh

Arid

Mobile sand
Avalanche faces

◄———— Wind

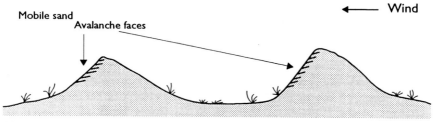

Vegetation on dune plinths only

Sub-arid to semi-arid

Dune crests mobile
Small avalanche faces

Pedogenic calcrete
Well developed vegetation cover

Semi-arid

Dunes stabilized by vegetation

Marsh and lacustrine deposition

Runoff   Runoff

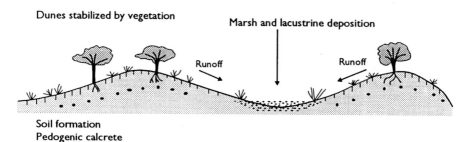

Soil formation
Pedogenic calcrete

Figure 8.1 Possible responses of dunes to changes in climate (after Rognon 1982).

deposits in interdune areas. In periods of intense rainfall, runoff may even erode dunes (Talbot and Williams 1978; Yair 1990). By contrast, periods of aridity will give rise to a very unfavourable biotic environment within the sand sea, with low water tables and active sand movement preventing the growth of vegetation.

DUNE SYSTEMS AND PALAEOCLIMATES

Evidence indicating that aeolian activity has been both more extensive and/or intense than it is at present is widely distributed in currently arid and semi-arid regions of the world. In addition to the effects of the major climatic changes of the Pleistocene, there is an increasing body of evidence that demonstrates the dynamic nature of many dune systems, many of which display multiple periods of dune stabilization and reactivation in the late Holocene.

Dune systems that are stabilized or fixed by vegetation and subsequent soil formation may provide valuable information on the former extent of aridity and on the nature of past wind regimes and regional circulation patterns, but can only be used to reconstruct palaeoclimates with confidence if periods of dune activity can be dated reliably and the relations between modern winds and dune alignments are known.

Dune activity classifications

Considerable confusion has arisen in the debate about the relations between dunes and past climates because of the problems of definition of dunes as 'active', 'inactive', 'mobile' or 'stable'. Part of the problem is that 'active' and 'inactive' dunes represent end members of a continuum that ranges from mobile (rapidly migrating) active barchan dunes to linear dunes that are stabilized by soil formation or savanna woodland, and climbing and falling dunes covered by colluvial mantles (Thomas and Shaw 1991a; Thomas 1992). Any definition must also take into account that dunes are one example of a variety of aeolian depositional landforms that range from sand sheets to climbing and falling dunes.

Dunes can be classified as active, dormant, or relict based upon geomorphic and sedimentary criteria. Active aeolian landforms are features on which contemporary surface sand transport and deposition takes place, leading to the development of wind-rippled surfaces and, locally, avalanche faces at the angle of repose (Plate 22a). Primary sedimentary structures (*sensu* Hunter (1977)) are preserved. Depending on their morphological type, active dunes may be migrating in the net transport direction (crescentic or transverse dunes), extending (e.g. linear dunes), or vertically accreting (e.g. star dunes). Vegetation may be present to varying degrees and may be permanent, seasonal, or ephemeral. The degree of aeolian activity may vary seasonally, annually or decadally in response to changes in sand supply, wind velocity, vegetation cover, and moisture content.

Dormant aeolian landforms (Plate 22b) are those on which surface sand transport and deposition are absent or at a low level. Wind-rippled surfaces are rare or absent, as are avalanche faces at the angle of repose. Old avalanche slopes may be degraded, so that lee face slope angles of 20° or less are common on crescentic dunes. Primary sedimentary structures are still present, but may be partially destroyed by bioturbation (Lancaster 1992). Vegetation cover is usually well developed and includes a high percentage of perennial plants, including shrubs and locally trees (Bowers 1982). The sand surface may be stabilized with biogenic crusts. Sand mobility is reduced by a low energy wind regime, lack of sand supply, and/or presence of a well-developed vegetation cover. Dormant dunes may revert to active features as a result of cyclic or secular environmental changes. A wide variety of states of dune dormancy can occur.

Relict aeolian landforms (Plate 22c) are those that are clearly a product of past climatic regimes or depositional environments and have been stabilized for some considerable period of time. They include dunes and sand sheets that are stabilized by soil development (including calcic horizons) (Vökel and Grunert 1990), early stages of diagenesis (partial cementation) (Talbot 1985), deflation lag surfaces, colluvial cover (Tchakerian 1992), and woodland vegetation (Grove 1969). Relict features may revert to an active state only as a result of major environmental changes.

Dating of periods of dune formation and stabilization

Until recently, a major problem in using aeolian deposits and landforms to reconstruct palaeoclimates was the difficulty of dating episodes of dune formation and aeolian deposition. Reliable dating of periods of dune activity is critical, as many dune fields indicate multiple episodes of dune formation. In some areas, ^{14}C dating of organic materials in palaeosols can be used to bracket periods of aeolian deposition (e.g. Forman and Maat 1990; Wells et al. 1990; Kocurek et al. 1991), but the majority of desert aeolian depositional settings do not favour formation and/or preservation of such materials.

Luminescence dating by thermoluminescence (TL) and optically stimulated luminescence (OSL) techniques provides the means to establish a chronology of periods of aeolian activity and dune formation by measuring the time since burial of sediments and/or stabilization of areas of dunes (Wintle and Huntley 1982; Berger 1988; Wintle 1993). Luminescence dating has been applied successfully to develop chronologies of dune formation and reactivation in Australia (e.g. Gardner et al. 1987; Chen et al. 1990; Lees et al. 1990; Nanson et al. 1992) and North America (e.g. Forman and Maat 1990; Stokes 1991; Edwards 1993). Relative dating schemes based on pedological and sedimentological criteria as well as stratigraphic evidence can

(a)

(b)

(c)

Plate 22 Examples of relict, dormant, and active dunes at Kelso Dunes, California:
(a) Active dunes, (b) Dormant dunes (last active 150 yr ago), and (c) Relict dunes
stabilized 3.5 to 4 ka).

then be used to develop local and regional histories of dune-forming episodes
(e.g. Tchakerian 1992). The availability of absolute dates for periods of
aeolian deposition also facilitates correlation with palaeoclimatic data from
other sources, and leads to more appropriate palaeoclimatic reconstructions.

Relict dune systems

Extensive systems of relict dunes, many of which are covered by savanna or
grassland vegetation, have been recognized from the margins of modern
active sand seas in the Sahel (e.g. Grove and Warren 1968; Talbot 1980, 1984),
southern Africa (Lancaster 1981a; Thomas 1984), India (Wasson *et al.* 1983;
Chawla *et al.* 1991), Australia (Bowler 1976; Wasson 1984), and throughout
the western United States (Ahlbrandt *et al.* 1983; Holliday 1989; Gaylord
1990). Relict dune systems have been identified from aerial photographs and
satellite images (Plate 23) in areas with up to 1,000 mm of annual precipitation, several hundred kilometres from the margin of areas of active dunes.
    In North America, currently vegetation-stabilized (? dormant) dunes on

Plate 23 Landsat image of relict linear dunes in the western Kalahari in Namibia.

the Great Plains were active after 9 ka with maximum aeolian erosion and deposition 6 to 4.5 ka (Holliday 1989). Similar chronologies are recognized in Wyoming (Gaylord 1990) with at least four episodes of enhanced aeolian activity after 7.5 ka: maximum aridity occurred 7.5 to 7.0 ka with a lesser peak 5.94 to 4.54 ka. The Nebraska Sand Hills have experienced several periods of reactivation and/or dune formation in the Holocene (Ahlbrandt *et al.* 1983). In many areas of the Great Plains, the most recent dune activity occurred between 3.5 to 1.5 ka (see summary in Muhs and Maat (1993)), but evidence for aeolian activity in the past 1,000 yr is now becoming available (e.g. Madole 1994). All periods of dune activity occurred during warmer and drier episodes, with prolonged drought between 6.5 and 4.5 ka. Periods of dune activity in the southwestern United States are less well constrained, but Stokes and Breed (1993) recognize three periods of dune reactivation 0.4, 2 to 3 ka and 4.7 ka. In the Mojave Desert, major periods of aeolian activity occurred between 30 and 20 ka and 15 to 7 ka, with significant periods of Holocene activity around 4 ka, 2.3 to 1.4, 0.8 to 0.4, and 0.2 ka (Wintle *et al.* 1994; Clarke *et al.* in press; Rendell in press).

In the Sahel and southern Sahara, three main dune generations are recognized (Grove and Warren 1968; Talbot 1980). The oldest dunes are very degraded and were formed prior to 20 ka. A subsequent period of dune formation and/or remobilization occurred during the period 13–20 ka and affected a very wide area (Talbot 1980). Most of the dunes were stabilized by vegetation and soil formation during the early Holocene, 7 to 11 ka (Talbot 1985; Vökel and Grunert 1990) during a period of humid conditions throughout the region. The third period of dune reactivation occurred after 5 ka. In the past several decades, regional drought and vegetation destruction by human activities have resulted in the remobilization of many dunes and even formation of new dune areas (Nickling 1993).

Extensive systems of dunes (Plate 23, Figure 8.2), mostly of linear form, occur throughout the Kalahari from the Orange River at 28° S to latitude 12° S in Angola and western Zambia. Today, these dunes are stabilized by savanna vegetation and are found in areas where rainfall is up to 800 mm per year (Grove 1969; Lancaster 1981a; Thomas 1984). The dunes form a massive semi-circular arc with a radius of some 1,000 km which corresponds approximately to the pattern of winds outblowing around the southern African anticyclone situated over the northern Transvaal. Within this arc, three distinct subsystems of dunes can be identified. The northern subsystem consists of dunes on E–W or ENE–WSW alignments. North and west of the Okavango Delta, they consist of broad straight parallel ridges up to 25 m high and 200 km long, with a spacing of 1.5–3 km. The dunes are very degraded and support a dense cover of tree savanna and open savanna woodland. In western Zimbabwe and adjacent areas of Botswana, the second subsystem consists of dunes with a relief of up to 25 m and 1.5–2 km apart (Flint and Bond 1968; Thomas 1984). The southern dunes form the third subsystem and are best developed in a 100–200 km wide belt between the highlands of Namibia and the Orange River near Upington. They consist of NNW–SSE or WNW–ESE trending parallel to sub-parallel ridges 5–15 m high with a spacing of 0.2–0.4 km, with their steeper slopes facing southwest (Lewis 1936; Lancaster 1988b).

The ages of the dune systems are presently unknown. The very large dunes of the northern Kalahari and adjacent areas of Zaire, Angola, and Zambia have been suggested to be of late Pliocene age (Partridge 1993), whereas linear dunes of the southwestern Kalahari may have been formed or reformed in the late Pleistocene (Lancaster 1989c). Thomas and Shaw (1991b) and Livingstone and Thomas (1993) have, however, questioned the relict status and therefore the palaeoclimatic significance of these dunes, and it appears that many dunes in the southwestern Kalahari appear to be episodically active today.

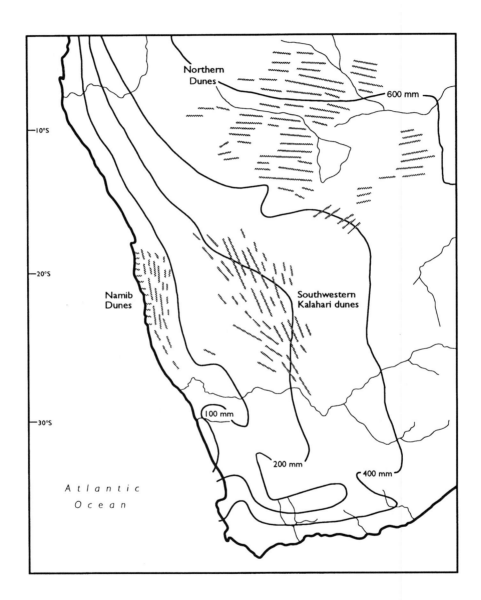

Figure 8.2 Relict and dormant dune systems in the Kalahari region (after Lancaster 1981a).

PALAEOCLIMATIC INFORMATION FROM DUNE SYSTEMS

Changes in the extent of arid conditions

Active continental dunes occur today, where sand supply conditions permit, in areas where mean annual rainfall is 100–150 mm or less. The existence of such dunes has long been regarded as an unequivocal index of arid climates. Comparison of the precipitation limits of active and fixed dunes can therefore provide an indication of the magnitude of changes in the position and extent of climatic belts (Sarnthein 1978). This approach has proved useful in the Sahel zone of western Africa, where there is a strong latitudinal pattern to climatic and vegetation zones (Figure 8.3). However, in the eastern Sahara, Australia, and southern Africa the evidence suggests that the area of aridity may have expanded and contracted about a central core region which has always remained arid (Haynes 1982; Lancaster 1984).

Recently, assumptions regarding the palaeoclimatic significance of vegetated dune systems have been challenged (Thomas and Shaw 1991b; Thomas 1992). It is now clear that interpretations of the distribution of active, dormant, and relict dunes in terms of changes in climatic zones are oversimplified. The importance of interactions between wind velocity and vegetation cover (and hence rainfall and soil moisture) on sand mobility are now recognized in Australia (Ash and Wasson 1983), the Sahel (Talbot 1984), southern Africa (Lancaster 1988b; Thomas and Tsoar 1990; Livingstone and Thomas 1993), and the western United States (Muhs and Maat 1993; Wintle et al. 1994). Periods of reactivation or formation of dunes may therefore be the result of changes in wind velocity and/or the moisture balance (i.e.

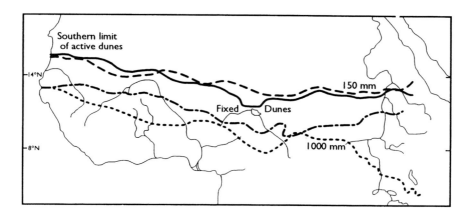

Figure 8.3 Limits of active and relict dunes in the Sahel region of Africa (after Lancaster 1990).

intensification of aridity). However in some areas, enhanced sand supply from desiccation of lake basins or fluvial systems is a major control on the timing of periods of dune formation (e.g. Lancaster 1993b).

Changes in wind velocity and/or moisture balance

The mobility of desert dunes is directly proportional to the sand-moving power of the wind, but inversely proportional to their vegetation cover (Ash and Wasson 1983). In turn, vegetation cover is a function of the ratio between annual rainfall (P) and potential evapotranspiration (PE). An index of the sand-moving power of the wind is the percentage of the time the wind is blowing above the threshold velocity for sand transport (4.5 m/second) (W). The ratio between these two terms gives a mobility index (Lancaster 1988b):

$$M = W / (P/PE) \qquad (8.1)$$

Threshold values for dune mobility determined from field observations of linear dunes in southern Africa as well as sand sheets, crescentic, and parabolic dunes in the Great Plains of the United States (Figure 8.4), suggest that dunes are fully active when values for M exceed 200 and most dunes are completely stabilized by vegetation when M is less than 50 (Lancaster 1988b; Muhs and Maat 1993).

In southern Africa, there is a gradient in sand mobility, as defined by this relationship, from the northern and eastern parts of the southwestern Kalahari to the active Namib Sand Sea (Figure 8.4a), that accurately parallels observations of the amounts of sand movement on dunes in the region (Lancaster 1988b). Significant increases in wind velocity or changes in the moisture balance must have occurred during periods when the dunes in the southwestern Kalahari were formed or remobilized. Recent studies of dune dynamics in the southwestern Kalahari suggest however that a considerable amount of sediment transport takes place on these dunes in present climatic conditions. This indicates that relatively minor changes in effective precipitation and vegetation cover could result in major changes in dune activity in this region (Livingstone and Thomas 1993).

In the Sahel, Talbot (1984) concluded that, despite an increase in late Pleistocene wind velocities of 50 per cent, dune formation and reactivation in the period 25 or 20 ka to 13 ka would have required a reduction in rainfall to 25–50 per cent present values. Ash and Wasson (1983) suggested that an increase in wind velocity of 20–30 per cent would result in the mobilization of Simpson Desert linear dunes. Wasson (1984) argued that increased windiness during the late Pleistocene dune-building phase (20 to 13 ka) would have increased PE and enhanced aridity, even without any reduction in rainfall.

In the Mojave Desert, preliminary assessments of the limits to dune activity suggest that the current seasonal distribution of precipitation and

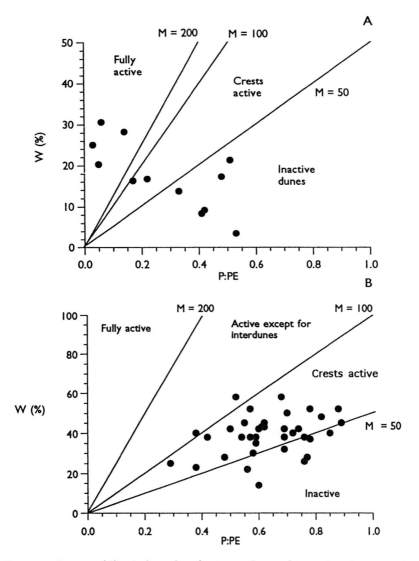

Figure 8.4 Dune mobility index values for A: southern Africa and B: the Great Plains of the USA (modified from Muhs and Maat 1993).

wind speeds in the region is unfavourable for active sand transport and dune formation. Sand transport potential is greatest in late winter and spring, when seasonal or ephemeral vegetation cover is at a maximum. Reduced winter precipitation and increased summer temperatures, as have been suggested for this area in the mid-Holocene (Spaulding 1991), would probably increase aeolian activity in areas where sediment supply is not a limiting factor. Values

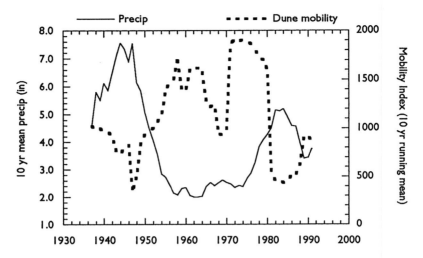

Figure 8.5 Relations between 10-year running means of annual precipitation and dune mobility index at Palm Springs, California.

of *M* for areas adjacent to Kelso Dunes are strongly influenced by changes in rainfall from year to year, and range from 100 (1984–1989 mean) to 131 (1951–1960 mean), indicating that only dune crests should be active today, a prediction supported by field observations. Assuming no change in wind speeds, simulations of the mobility index indicate that rainfall must have decreased to 50 per cent of modern mean values (~ 76 mm yr^{-1}) during periods of aeolian activity (Lancaster 1993b).

Sensitivity analyses of the dune mobility index for parts of the southern California deserts suggest that it is strongly dependent on annual precipitation amounts. As on the Great Plains (Muhs and Maat 1993), wind energy in this region is always sufficient for active sand transport. Over the past several decades, however, dune mobility has varied widely. Ten-year running means of the dune mobility and precipitation for the Palm Springs area (Figure 8.5) show a strong correlation with both the area of active dunes and dune migration rates in the Coachella Valley, suggesting that in this area annual precipitation has a major effect on dune mobility by promoting changes in the vegetation cover.

Comparisons between areas of different dunes suggest that the critical factor controlling dune mobility varies from one region to another. It appears to be wind strength in the Australian and Sonoran deserts, but the seasonal distribution and/or annual total amount of precipitation in the Kalahari, the Great Plains, the Mojave, and southern California regions.

Palaeowinds and circulation patterns

Comparisons between the modern pattern of sand-moving winds and dune alignments may provide important information on palaeocirculation patterns. In some areas, the alignments of relict dunes depart significantly from the direction of modern sand-moving winds, suggesting that when they were formed, wind regimes differed from those found today (e.g. Brookfield 1970; Lancaster 1981a; Wells 1983; Muhs 1985; Lancaster 1993b). In other regions, such as the southern Sahara and Sahel, dune alignment patterns indicate that most relict dunes were formed in wind regimes similar to those occurring today (Mainguet and Canon 1976; Talbot 1980), with a dominance of easterly to northeasterly winds. Fryberger (1980) however provides evidence from crossing trends of dunes in Mauritania, Mali, and Niger of changes in wind directions between the oldest dunes that were formed by more easterly winds and those formed 13–20 ka by winds similar to the northeasterlies occurring today.

Such conclusions are only valid if: (1) there is a good knowledge of the characteristics of the modern wind regime; (2) the relations between dune alignments and winds are known from studies of active dunes; and (3) formation of different areas of dunes can be assigned to a specific time period (Thomas and Shaw 1991b). Inferences of palaeowind directions from dunes are most reliable for areas of crescentic and parabolic dunes that form in simple wind regimes (Marrs and Kolm 1982; Wells 1983; Muhs 1985). Small dunes of these types have short reconstitution times and can be entirely reworked to reflect changed wind directions in periods of decades. Their alignment patterns therefore directly reflect the wind directions at the time when they were formed, rather than multiple episodes of reworking and modification of pre-existing forms. In addition, it is necessary to be able to reliably assign often widely separated dune areas to a specific time period (e.g. by absolute dating) in order to make appropriate correlation between dune patterns and formative winds.

In the Kalahari region of southern Africa, Lancaster (1981a) and Thomas (1984) have recognized three systems of fixed dunes, each corresponding to a different palaeowind pattern. Patterns of winds for all periods of dune formation (Figure 8.6) suggest that periods of aridity and dune formation in southern Africa were associated with, and probably caused by, increased persistence and strength of anticyclonic circulations, which also prevented moisture-bearing winds from reaching the interior of the subcontinent.

In North America, Wells (1983) compared modern winds and the alignments of dunes supposedly formed during a late Glacial period of aeolian activity that peaked 14 ka, and inferred a shift in wind directions of 40–90° towards the north. The change in wind directions was especially marked in the southern Great Plains, and least in the desert regions of the southwest. Wells attributed the changed wind regime to the development of

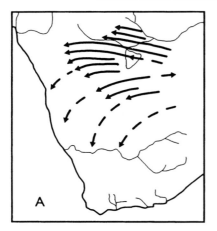

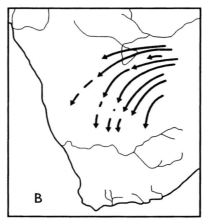

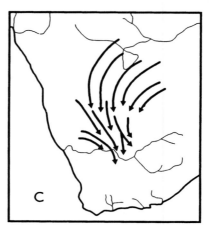

Figure 8.6 Inferred dune-forming
circulation patterns for Kalahari relict
dunes (after Lancaster 1981a).

three Rossby waves at latitudes 40 to 50° N, whereas the modern circulation typically displays a 4 or 5 wave pattern.

In the Mojave Desert, many relict dune patterns suggest an increased frequency of northerly winds. Preliminary palaeoclimatic models for this region suggest that increased sand transport from northerly directions is associated with more frequent and/or intense winter frontal systems. Therefore dunes aligned with respect to northerly winds imply more frequent and/or intense winter aeolian transport events, whereas dunes aligned with westerly winds indicate dominance of spring and early summer circulations (Lancaster 1990).

In Australia, linear dune alignments parallel modern resultant sand-transport directions in northern and eastern parts of the arid zone (Bowler and Wasson 1984). However, in the southern Strezlecki desert, and the semi-arid Mallee area, dune alignments are W–E, but modern winds are from the southwest (Sprigg 1979, 1982; Bowler and Wasson 1984). Increased frequency of summer west and northwest winds during the period of dune formation 12 to 20 ka was attributed by Bowler and Wasson (1984) to steeper pressure gradients around the anticyclonic belts and more frequent passage of fronts to the south of the continent.

INTERDUNE DEPOSITS

Interdune deposits are widespread in many modern sand seas (e.g. Ahlbrandt and Fryberger 1981; Lancaster and Teller 1988). Such deposits may be formed in shallow lakes, incursion of fluvial systems into dune areas, or by precipitation from shallow groundwater. Interdune lacustrine deposits and their associated fossil faunas can provide data on the extent and nature of humid conditions within and adjacent to the sand sea; and the chronology of sand sea accumulation through radiometric dating and studies of archaeological material. Interdune deposits that formed as the result of the penetration of extra-dune fluvial environments into the sand sea can provide data on humid conditions in headwater regions adjacent to the sand sea, as well as the development of dune patterns.

In the Namib Sand Sea, interdune lacustrine deposits of restricted extent have been laid down periodically during the accumulation of the sand sea (Figure 8.7). These carbonate-rich deposits are indicative of increased moisture availability in this normally hyper-arid to arid region. Diatom and mollusk faunas indicate that the water bodies were fresh to brackish in composition (Teller et al. 1990). Modern analogues suggest several possible depositional environments. Shallow seasonal or ephemeral lakes may have formed in interdune areas as a result of increased regional or local precipitation. Long-continued periods of increased rainfall may have raised local groundwater levels such that seepage occurred from shallow aquifers. In some areas on the northern margin of the Sand Sea adjacent to the Kuiseb

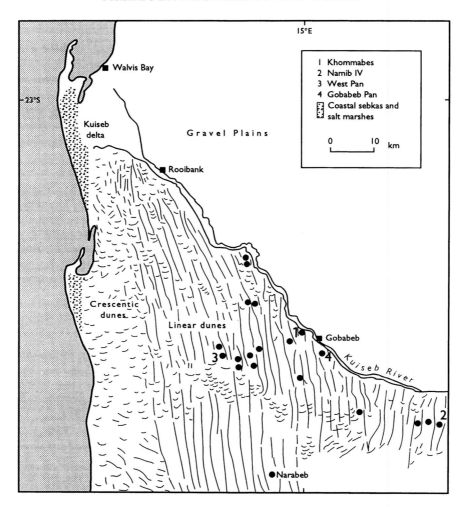

Figure 8.7 Distribution of interdune lacustrine deposits in the northern part of the Namib Sand Sea (after Teller *et al.* 1990).

River, lakes may have formed adjacent to ephemeral water courses in response to high groundwater levels in those river valleys. At other localities, for example at Narabeb, lakes formed at the former end points of ephemeral rivers, which at one time were able to penetrate much farther west into the Sand Sea than they do now (Seely and Sandelowsky 1974; Teller and Lancaster 1986). This implies that rainfall and runoff in the headwaters of the rivers in the highlands east of the Sand Sea were at times greater and/or that linear dune patterns were sufficiently open to permit water to flow as much as 40 km west of the present terminal playas of these rivers. Radiocarbon dates on many of the interdune lacustrine carbonates and molluscs cluster between 20 and 32 ka (Vogel 1989; Teller *et al.* 1990). Evidence from elsewhere in the arid zone of southern Africa (Deacon and Lancaster 1988) indicates that rainfall throughout this climatic zone was probably significantly higher during the period when these lacustrine beds were deposited.

In northern Saharan sand seas, increased flooding of the Wadi Saoura and lakes in the Chott Djerid basin also indicate increased moisture availability between 40 and 20 ka (Rognon 1987). Similarly wet conditions occurred in the southern Sahara sand seas during the period 30 to 20 ka (Talbot, 1980; Fabre and Petit-Maire 1988) and again in the Holocene between 9 and 7 ka, when lacustrine conditions were very extensive (Rognon and Williams 1977; Petit-Maire *et al.* 1987; Fabre and Petit-Maire 1988). Periods of increased moisture are indicated for the area of the Murzuq sand sea of Libya at 160 to 125, 90, and 26 to 22.5 ka (Petit-Maire *et al.* 1980). In the Great Sand Sea of western Egypt, Haynes (1982) reports periods of increased rainfall in the Middle Palaeolithic and Neolithic. The extent of the Neolithic 'pluvial' is confirmed by pollen evidence that indicates permanent lakes surrounded by savanna woodland in the now hyper-arid northwest Sudan between 8.5 and 6.1 ka (e.g. Ritchie *et al.* 1985; Haynes and Mead 1987), and by a variety of evidence for high lake levels throughout the southern Saharan margins (e.g. Street and Grove 1979). In Saudi Arabia, interdune lacustrine deposits from the Nafud and Rub' al Khali sand seas indicate moist conditions between 3.2 and 2.4 ka and again between 8.5 and 5 ka (Whitney *et al.* 1983).

EFFECTS OF CLIMATIC CHANGES ON SAND SEA AND DUNE DEVELOPMENT

There is an increasing body of evidence that suggests that many modern sand seas have accumulated episodically, with Quaternary climatic and sea level changes exerting a major control on aeolian sedimentation patterns in space and time. The legacy of past wind regimes, in the form of dune patterns out of equilibrium with modern winds, has been frequently cited as an example of the effects of such climatic changes (e.g. Glennie 1970; Besler 1980). However, recent work suggests that a major manifestation of the effects of climatic changes on sand sea development is the formation of different generations of dunes and different periods of deposition in complex and compound dunes. The key to understanding the episodic nature of aeolian accumulation has been the recognition that major regionally-extensive bounding surfaces (super surfaces) separate genetic units in the deposits of many ancient and modern sand seas (Kocurek 1988).

Super surfaces (Figure 8.8) truncate aeolian deposits and therefore represent periods when aeolian accumulation ceased, to be replaced by sediment

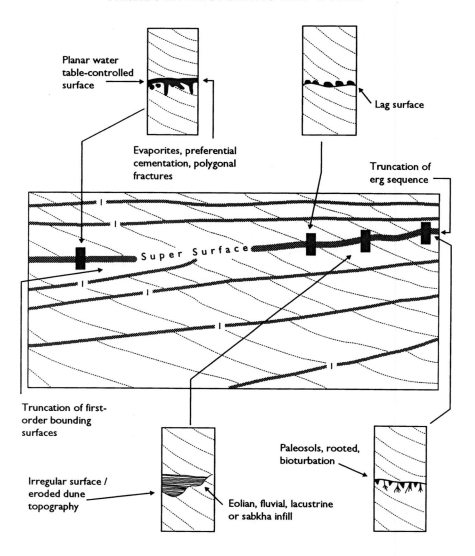

Planar water table-controlled surface

Evaporites, preferential cementation, polygonal fractures

Lag surface

Truncation of erg sequence

Super Surface

Truncation of first-order bounding surfaces

Irregular surface / eroded dune topography

Eolian, fluvial, lacustrine or sabkha infill

Paleosols, rooted, bioturbation

Figure 8.8 Features associated with regionally extensive bounding surfaces (super surfaces) (after Kocurek 1988).

bypassing, geomorphic stability or deflation over part or all of the area of a sand sea. Aeolian accumulation implies a positive sediment budget for an area. Super surface development will occur when the sediment budget for the area is neutral (output of sediment = input) forming a bypass super surface; when the sediment budget is negative (output of sediment > input) leading to an erosional super surface; or when the dunes are stabilized by vegetation (stabilized super surface) (Kocurek and Havholm 1994).

Changes in sand supply

Many sand seas are fed by material deflated from areas of marine, fluvial or lacustrine sands. High deflation rates are dependent upon active re-supply of the source areas as protective lags form rapidly in areas of mixed sediments

Figure 8.9 Relations between linear dune systems and fluvial source zones in the Simpson Desert of Australia (after Wasson 1983).

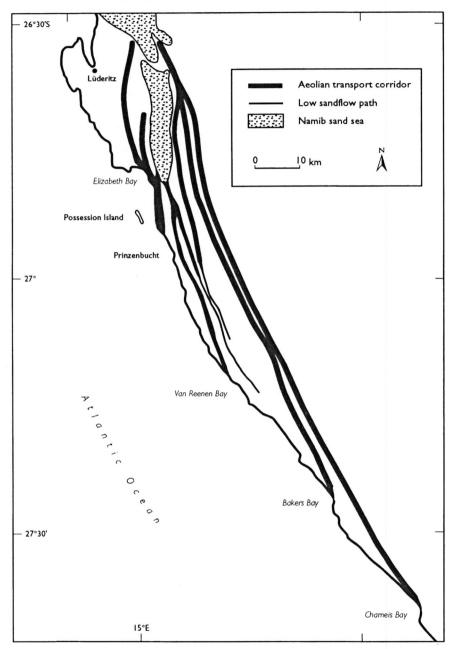

Figure 8.10 Modern and past positions of sand transport corridors from coastal source zones to the Namib Sand Sea (after Corbett 1993).

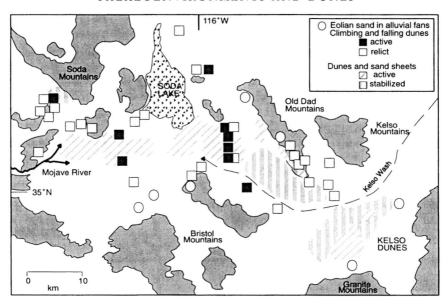

Figure 8.11 Modern and relict aeolian deposits in the Kelso Dunes area, Mojave Desert (after Lancaster 1994b).

(Wilson 1971). There is thus a close link between dune activity and the supply of sediment so that dune formation may be favoured in periods of geomorphic instability and enhanced fluvial sediment loads. In addition, a high sediment supply to dunes promotes sand transport close to saturation levels and rapid deposition and dune migration. It also tends to limit colonization by annual plants and the regrowth of perennials (Pye and Tsoar 1990).

In the Simpson-Strzelecki desert of Australia, Wasson (1983) has demonstrated the close links that exist between linear dune formation and sediment supply from late Pleistocene floodplain sediments (Figure 8.9), with clay pellet formation promoted by high saline groundwater levels. Reduction of sediment loads following late Glacial climatic changes effectively terminated sediment input to the Great Sand Dunes (Johnson 1967). Littoral sands exposed during Glacial low-stands also provided a ready source for dunes in the Namib, southern Australia (Sprigg 1979) and the Arabian Peninsula (Shinn 1973; Warren 1988). In the southern Namib, changes in sea levels gave rise to a different coastal configuration and thus to shifts in the position of the sand transport corridors that feed the sand sea (Figure 8.10) (Corbett 1993).

Evidence indicating that aeolian activity has been both more extensive and more intense than it is at present occurs throughout the Mojave Desert

(Figure 8.11) (e.g. Smith 1967; Tchakerian 1992). Major sources of palaeoclimatic information are dormant vegetation-stabilized dune systems and relict sand ramps against mountain fronts. The sand ramps represent the amalgamation of deposits of sand sheets and climbing and falling dunes, talus derived from adjacent mountain slopes, as well as fluvial and colluvial sediments (Tchakerian 1992). Two major regionally-extensive aeolian depositional episodes have been identified in the past 40 ka using luminescence dating (Rendell in press): (1) 30–20 ka, and (2) 15 ka–7 ka. Significant periods of Holocene aeolian activity also occurred in the Mojave River–Kelso Dunes sediment transport system at around 4 ka, 2.3–1.4 ka, 0.8–0.4 ka, and 0.2 ka (Wintle *et al.* 1994; Clarke *et al.* in press).

It appears that there was a major change in the aeolian environment in the Mojave Desert during the early Holocene (Rendell *et al.* in press). Prior to this time, sediment supply from fluctuating pluvial lakes (e.g. Lake Mojave) and more active fluvial systems was sufficient to promote the accumulation

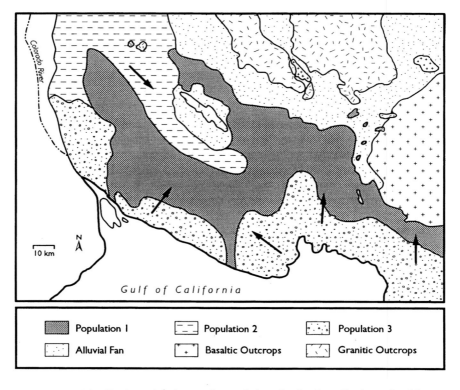

Figure 8.12 Distribution of major sand populations in the Gran Desierto Sand Sea as a reflection of changes in sediment supply. Arrows indicate former transport directions.

of large climbing and falling dunes. Each period of lake desiccation provided a ready sediment supply for aeolian processes in the region. After the desiccation of the region in the latest Pleistocene and early Holocene, sediment supply declined dramatically so that later Holocene aeolian activity was restricted to the reworking of major dune fields (e.g. Kelso) and to areas of active sediment supply (e.g. the Cronese Basin).

In the Gran Desierto (Figure 8.12), eustatic changes in sea level have played a major role in determining sand supply to dunes. Currently, coastal source areas are blocked by extensive salt marshes. Increased sediment input from these sources implies greater availability of sediment from this area, and changes in the configuration of the coastline to permit onshore transport of sand by the wind. For example, during the last glacial maximum, when sea level was ~ 120 m lower than present 21 ka (Bard *et al.* 1990), the coastline of the Bahia del Adair was located as much as 150 km south of its modern position and lay close to the position of the Colorado River delta front at this time (van Andel 1964), exposing a 50–100 km wide zone of sandy sediments. These conditions would have provided a ready source of material for input to the sand sea by southerly winds.

The primary sand supply for dunes in the western part of the sand sea appears to have been from point bars and river terrace deposits of the lower Colorado River downstream from Yuma (Blount and Lancaster 1990). The bedload of the lower Colorado River contains 30–40 per cent sand-sized particles, with modal sizes prior to dam construction between 250 and 125 µm (Thompson 1968), or similar to those in most dune sands. The occurrence of coarse sand in the northwestern parts of the sand sea suggests derivation from deposits laid down when the river had a higher competence than in historical times.

Fluctuations in sand supply from the Colorado River source may have been the result of eustatic changes in sea level or tectonic events that altered the pattern of fluvial sedimentation in the delta system. For example, during the last glacial maximum the delta front was located 100 km southeast of its modern position (van Andel 1964), resulting in formation of an incised valley in the vicinity of the present delta region. In this case, eustatic changes in sea level that affected fluvial and deltaic sand sources may have resulted in a switch from northwestern to southerly sand input to the Gran Desierto and the formation of entirely different dune areas.

A further control on sand supply to the Gran Desierto from the Colorado River is the periodic diversion of the river to flow into the Salton depression for periods of decades to centuries to form Lake Cahuilla (Waters 1983). Such diversions had a major effect on sediment supply to the Algodones Dunes (Muhs, in press). They likely cut off sand supply to the Gran Desierto sand sea. The avulsion of the Colorado River delta to the west following uplift of the Mesa Arenosa in the middle Pleistocene (Colletta and Ortlieb 1984) was also significant. Deflation of sediments from the abandoned floodplain could have provided the oldest population of sand in the sand sea (Blount and Lancaster 1990).

Effects of climatic changes on dune development

Recent stratigraphic and dating studies are demonstrating that many large dunes are composite features composed of multiple generations of sediments (e.g. Chawla *et al.* 1991; Rendell *et al.* 1993; Stokes and Breed 1993). The dunes may have cores that are tens of thousands of years old, with crestal areas that have been episodically active.

In Australia, the cores of linear dunes in the Simpson Desert appear to be composed of sediments deposited 80 ka ago or earlier, with subsequent periods of reworking and addition of material (Nanson *et al.* 1992). There was a major period of dune formation during the last glacial maximum (Wasson 1983; Nanson *et al.* 1992), with intermittent reworking of dune sediments during the Holocene.

Detailed studies of dune and interdune sediments in the Akchar Sand Sea of Mauritania by Kocurek *et al.* (1991) show that they represent the amalgamation of Late Pleistocene and Holocene deposits. The prominent large complex linear dunes (Figure 8.13) are composite features. Their core consists of sand deposited during the period 20 to 13 ka. These linear dunes were stabilized by vegetation during a period of increased rainfall 11 to 4.5 ka, when soil formation altered dune sediments and lakes formed in interdune areas. Further periods of dune formation after 4 ka cannibalized existing aeolian deposits on the upwind margin of the sand sea. The currently active 'cap' of crescentic dunes superimposed on the linear dunes dates to last 30 yr. Similar sequences of dune deposits have been recognized in the Sahel (Talbot 1985) and southern Sahara (Vökel and Grunert 1990).

In the Rub' al Khali, Nafud and Wahiba sand seas of the Arabian Peninsula aridity and dune formation occurred between 20 and 9 ka (McClure 1978; Whitney *et al.* 1983; Warren 1988). Linear and crescentic dunes were stabilized by vegetation with weak pedogenesis and lakes in interdune areas 9 to 6 ka. The morphology of the modern dune system may be less than 2 ka old (Figure 8.14).

Figure 8.13 Periods of dune formation and stabilization in the Akchar sand sea of Mauritania (after Kocurek *et al.* 1991). I: Linear dunes are formed, or are active; II: Dune stabilization by vegetation and super surface formation, with lakes in interdune areas; III: Dune reactivation and drying of lakes; IV: Further revegetation of dunes and supersurface formation; V: Modern reactivation of dunes.

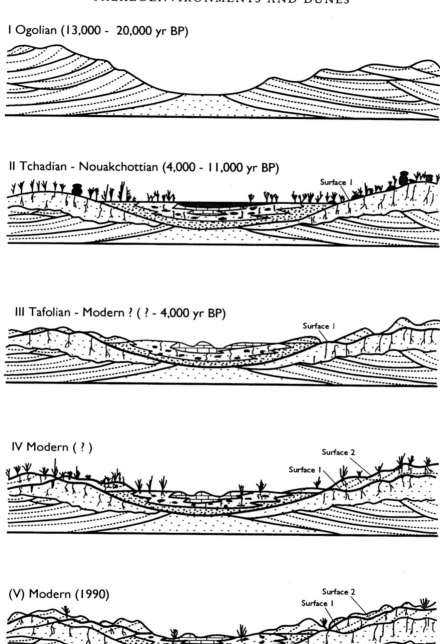

I Ogolian (13,000 - 20,000 yr BP)

II Tchadian - Nouakchottian (4,000 - 11,000 yr BP)

Surface 1

III Tafolian - Modern ? (? - 4,000 yr BP)

Surface 1

IV Modern (?)

Surface 2

Surface 1

(V) Modern (1990)

Surface 2

Surface 1

NW   SE

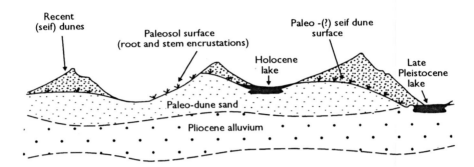

Figure 8.14 Inferred stratigraphic relations between periods of dune formation in the Rub' al Khali sand sea (after McClure 1978).

CONCLUSIONS

Sand seas are important sources of paleoclimatic information for arid regions and provide information on past wind regimes and the former extent and timing of arid conditions. Recent work on the climatic controls of dune mobility is providing better constraints on the nature of past episodes of dune formation and/or reactivation.

Sand sea accumulation has also been significantly affected by Quaternary climatic and sea level changes. Evidence from many major sand seas suggests that their accumulation has been episodic, with long periods of stability and reworking of dune sediments being interrupted by relatively brief episodes of sediment input and dune formation. Current aeolian activity in many areas is merely reworking and/or cannibalizing existing sediments leading to increasing complexity of the dune patterns. Stratigraphic and geomorphic studies in many areas are however suggesting the dynamic nature of aeolian activity during the Holocene with multiple periods of dune reactivation and stabilization on time scales of centuries to millenniums.

Observations suggesting that many large, complex dunes are the products of multiple episodes of accumulation and reworking of sediments have important implications for understanding the controls of dune morphology and the evolution of large dunes. They indicate that many of the larger dunes (especially linear and star dunes) in major sand seas may be composite forms with a long history and that the current dune forms and patterns may not reflect contemporary wind regime or sediment supply conditions.

9

REVIEW AND PROSPECT

This book documents the very significant advances in the level of under-standing of the geomorphology of desert dunes in the past two decades. This progress has been in three major areas: (1) regional studies of sand seas using satellite images have provided information on the environmental context and wind regimes of dunes so that discrete domains of different dune types can be identified; (2) detailed process studies on dunes have identified, and in many cases quantified, the major dune-forming processes and provided the physical basis for understanding why different dune types occur; and (3) stratigraphic and dating studies have begun to establish the history of sand seas and dune fields and to show how dunes have responded to climate change. In a sense, this book has documented the 'coming of age' of this field of geomorphology. Many of the questions posed in the introduction to this book can now be answered with a considerable degree of confidence. There remain however many areas where understanding is sketchy. These include in particular the controls of dune size and spacing, as well as the dynamics of dunes on decadal time scales and their response to environmental changes at all time scales.

Compared to many other branches of geomorphology, studies of desert dunes lack a well-developed conceptual framework. Away from the area of sand transport studies, the degree of understanding of the physical basis of processes is poor. Attempts to provide realistic simulation models for dune morphology and development have been few and mostly of questionable relevance. In part, these deficiencies are a consequence of the small number of active workers in the field and the difficulties of working in many desert areas. In many instances, collaboration between modellers and field inves-tigators has been minimal. Models are however necessary to point the way for field studies by guiding process measurements in the identification of critical parameters.

One reason that studies of desert dune geomorphology have lagged behind other branches of geomorphology (e.g. coastal processes and dynamics and fluvial processes) is that many dune areas are remote from major population centres and their impacts on human activities are limited. However, arid regions are home to a significant percentage of the world's population and include many areas of rapid economic expansion and population growth. The United Nations Environment Program has suggested that close to 1 billion people may be affected by desertification processes. A significant number of environmental problems in these areas involve dune processes. A major component of land degradation and desertification processes on desert margins is the reactivation of formerly vegetation-stabilized dunes and sand sheets (Mainguet 1985). Many dust storms are generated by the impact of saltating sand on silt and clay surfaces, and have major impacts on air quality in arid regions (e.g. Middleton *et al.* 1986; Nickling and Gillies 1993). Sand encroachment on highways, structures, and scarce irrigated agricultural land continues to be a major problem (e.g. Watson 1990; Khalaf *et al.* 1993).

Better understanding of dune processes and dynamics is essential for effective and appropriate identification and mitigation of geologic hazards in many arid regions (e.g. Tsoar 1990). For example, control and stabilization of active dunes with vegetation requires a good understanding of the degree of protection provided by different percentages of plant cover.

In some cases, knowledge of past dune dynamics can assist planning for future scenarios of environmental change. For example, studies of dune stratigraphy have indicated that many dune areas on the Great Plains of the United States were active in the past thousand years (Madole 1994). Future occurrence of such widespread periods of drought would have major impacts on groundwater resources, agricultural productivity, and air quality (via increased dust storm activity). As many areas of the Great Plains are close to climatic limits for active dunes, the regional response to future climatic changes can be modelled using the output from climate models to calculate dune mobility indices (Muhs and Maat 1993).

Although future progress in studies of desert dune geomorphology will undoubtedly involve fundamental investigations of processes and dynamics, it is clear, however, that understanding of environmental problems and geologic hazards in arid regions will increasingly drive the research effort. Such demands will undoubtedly identify further areas of incomplete understanding and stimulate new avenues for research.

CONCLUSIONS

This book has focused on dune processes and dynamics at different temporal and spatial scales, ranging from the micro-scale of sand transport processes, via meso-scale dune dynamics, to the regional or macro-scale and long periods of time involved in the development of dune patterns and sand seas. At each scale, however, variations in sand transport rates in time and space are the fundamental control of dune initiation, morphology, and accumulation.

It is clear that the factors that control the operation of the fundamental dune processes vary from one scale to the other. Sand transport is primarily determined by changes in wind shear velocity and turbulence with a period of tens of seconds or less. The morphology and dynamics of individual dunes appears to be controlled by seasonal changes in wind velocity and direction, but dunes may also respond to changes on a decadal scale. Process–form relations are most clearly identified at this scale. Large compound and complex dunes take centuries or millennia to change to form, as do many smaller dune fields. At this scale, and especially at the sand sea level, climatic changes are a major control on accumulation patterns in time and space and many features may be inherited from past environmental conditions. A major goal of present studies is to identify and distinguish between the effects of contemporary processes and past environments on present landforms.

Studies of desert dunes have advanced very significantly in recent years and have provided a new understanding of fundamental processes and new paradigms for research. The challenge for the future is to develop this understanding of process–form relations and to apply that knowledge to solution of environmental problems in arid lands.

REFERENCES

Ahlbrandt, T. S. 1974a 'The source of sand for the Killpecker Sand-Dune Field, southwestern Wyoming'. *Sedimentary Geology*, 11, 39–57.

—— 1974b 'Comparison of textures and structures to distinguish eolian environments, Kilpecker dune field, Wyoming'. *Mountain Geologist*, 12, 61–63.

—— 1979 'Textural parameters of eolian deposits'. In E. D. McKee (ed.), *A Study of Global Sand Seas*, 21–51, United States Geological Survey, Professional Paper 1052.

—— and Fryberger, S. G. 1980 *Eolian Deposits in the Nebraska Sand Hills*. United States Geological Survey, Professional Paper 1120-A.

—— and Fryberger, S. G. 1981 'Sedimentary features and significance of interdune deposits'. In F. G. Ethridge and R. M. Flores (eds), *Recent and Ancient Nonmarine Depositional Environments: Models for Exploration*, 293–314, Society of Economic Palaeontologists and Mineralogists, Tulsa, Oklahoma.

——, Swinehart, J. B. and Maroney, D. G. 1983 'The dynamic Holocene dune fields of the Great Plains and Rocky Mountain Basins, U.S.A.'. In M. E. Brookfield and T. S. Ahlbrandt (eds), *Eolian Sediments and Processes*, 379–406, Elsevier, Amsterdam.

Alimen, H. 1953 'Variations granulométriques et morphoscopiques de sable le long de profils dunaires au Sahara Occidental'. In *Actions Eoliennes*, 219–235, Colloques Internationaux, Centre National de Recherches Scientifiques, Paris.

——, Buron, M. and Chavaillon, J. 1958 'Caractères granulométriques des profils transversaux de quelques dunes d'Erg de Sahara Nord-occidental'. *Comptes Rendus de l'Académie des Sciences, Paris*, 247, 1758–1761.

Allen, J. R. L. 1968a *Current Ripples*, Elsevier, Amsterdam, 433 pp.

—— 1968b 'The nature and origin of bed-form hierarchies'. *Sedimentology*, 10, 161–182.

—— 1974 'Reaction, relaxation and lag in natural sedimentary systems: general principles, examples and lessons'. *Earth Science Reviews*, 10, 263–342.

—— 1984 *Sedimentary Structures: Their Character and Physical Basis*, Elsevier, Amsterdam, 593, 663 pp.

Allison, R. 1988 'Sediment types and sources in the Wahiba Sands'. In R. W. Dutton (ed.), *Scientific Results of the Royal Geographical Society's Oman Wahiba Sands Project 1985–1987*, 161–168, *Journal of Oman Studies*, Special Report 3, Muscat, Oman.

Anderson, R. S. 1987 'A theoretical model for aeolian impact ripples'. *Sedimentology*, 34, 943–956.

—— 1988 'The pattern of grainfall deposition in the lee of aeolian dunes'.

Sedimentology, 35, 175–188.

—— 1990 'Eolian ripples as examples of self-organization in geomorphological systems'. *Earth Science Reviews*, 29, 77–96.

—— and Haff, P. K. 1988 'Simulation of eolian saltation'. *Science*, 241, 820–823.

—— and Haff, P. K. 1991 'Wind modification and bed response during saltation of sand in air'. *Acta Mechanica*, Supplement 1, 21–52.

—— and Hallet, B. 1986 'Sediment transport by wind: toward a general model'. *Geological Society of America Bulletin*, 97, 523–535.

Andrews, S. 1981 'Sedimentology of Great Sand Dunes, Colorado'. In F. P. Ethridge and R. M. Flores (eds), *Recent and Ancient Non Marine Depositional Environments: Models for Exploration*, 279–291, The Society of Economic Paleontologists and Mineralogists, Tulsa, Oklahoma.

Angell, J. K., Pack, D. H. and Dickson, C. R. 1968 'A Langrangian study of helical circulations in the planetary boundary layer'. *Journal of Atmospheric Science*, 25, 707–717.

Anton, D. and Vincent, P. 1986 'Parabolic dunes of the Jafurah Desert, Eastern Province, Saudi Arabia'. *Journal of Arid Environments*, 11, 187–198.

Ash, J. E. and Wasson, R. J. 1983 'Vegetation and sand mobility in the Australian desert dunefield'. *Zeitschrift für Geomorphologie*, Supplement 45, 7–25.

Aufrère, L. 1928 'L'orientation des dunes et la direction des vents'. *Comptes Rendus de l'Académie des Sciences, Paris*, 187, 833–835.

Bagnold, R. A. 1933 'A further journey in the Libyan Desert'. *Geographical Journal*, 82, 103–129, 211–235.

—— 1941 *The Physics of Blown Sand and Desert Dunes*, Chapman and Hall, London, 265 pp.

—— 1951 'Sand formation in southern Arabia'. *Geographical Journal*, 117, 78–86.

—— 1953a 'The surface movement of blown sand in relation to meteorology', *Desert Research, Proceedings of the International Symposium, Research Council of Israel, Jerusalem*, 89–93.

—— 1953b 'Forme des dunes de sable et régime des vents'. In *Actions Eoliennes*, 23–32, Colloques Internationaux, Centre National de Recherches Scientifiques, Paris.

—— 1954 'Experiments on gravity-free dispersion of large solid spheres in a Newtonian fluid under shear'. *Proceedings of the Royal Society*, 225 A, 49–63.

—— 1956 'The flow of cohesionless grains in fluids'. *Proceedings of the Royal Society*, 249, 235–297.

Bard, E., Hamelin, B., Fairbanks, R. G. and Zindler, A. 1990 'Calibration of the ^{14}C timescale over the past 80,000 years using mass spectrometric U-Th ages from Barbados corals'. *Nature*, 345, 405–410.

Barndorff-Nielsen, O. and Christiansen, C. 1985 'The hyperbolic shape triangle and classification of sand sediments', *Proceedings of International Workshop on the Physics of Blown Sand*, Aarhus, University of Aarhus, 649–676.

——, Dalsgaard, K., Halcreen, C., Kuhlman, H., Møller, J. T. and Schou, G. 1982 'Variation in particle size distribution over a small dune'. *Sedimentology*, 29, 55–65.

Beadnell, H. J. L. 1910 'The sand dunes of the Libyan Desert'. *Geographical Journal*, 35, 379–395.

Bellair, P. 1953 *Sables desertiques et morphologie eolienne*, Proceedings of the 19th International Geological Congress, Algiers, 113–117.

Belly, Y. 1964 'Sand movement by wind'. *U.S. Army Corps of Engineers, Coastal Engineering Research Center, Technical Memo*, 1, Addendum III, 24 pp.

Berger, G. W. 1988 'Dating Quaternary events by luminescence'. In D. J. Easterbrook

REFERENCES

(ed.), *Dating Quaternary Sediments*, 13–50, Special Paper 227, Geological Society of America, Boulder, Colorado.

Besler, H. 1977 'Fluviale und aolische formung zwischen schott und erg'. *Stuttgarter Geographische Studien*, 91, 19–81.

—— 1980 'Die Dunen-Namib: entstehung und dynamik eines ergs'. *Stuttgarter Geographische Studien*, 96, 241 pp.

—— 1982 'The north-eastern Rub' al Khali within the borders of the United Arab Emirates'. *Zeitschrift für Geomorphology*, 26, 495–504.

—— and Marker, M. E. 1979 'Namib sandstone: a distinct lithological unit'. *Transactions of the Geological Society of South Africa*, 82, 155–160.

Blandford, W. T. 1876 'On the physical geography of the Great Indian Desert with especial reference to the existence of the sea in the Indus valley and on the origin and mode of formation of the sand hills'. *Journal of the Asiatic Society of Bengal (Calcutta)*, 45, 86–103.

Blount, G. and Lancaster, N. 1990 'Development of the Gran Desierto Sand Sea'. *Geology*, 18, 724–728.

Blumberg, D. G. and Greeley, R. 1993 'Field studies of aerodynamic roughness length'. *Journal of Arid Environments*, 25, 39–48.

Bowen, A. J. and Lindley, D. 1977 'A wind tunnel investigation of the wind speed and turbulence characteristics close to the ground over various escarpment shapes'. *Boundary Layer Meteorology*, 12, 259–271.

Bowers, J. 1982 'The plant ecology of inland dunes in western North America'. *Journal of Arid Environments*, 5, 199–220.

Bowler, J. M. 1976 'Aridity in Australia: age, origins and expression in aeolian landforms and sediments'. *Earth Science Reviews*, 12, 279–310.

—— 1978 'Glacial age aeolian events at high and low latitudes: a Southern Hemisphere perspective'. In E. M. Van Zinderen Bakker (ed.), *Antarctic Glacial History and World Palaeoenvironments*, 149–172, A. A. Balkema, Rotterdam.

—— 1983 'Lunettes as indices of hydrogeologic change: a review of Australian evidence'. *Proceedings of the Royal Society of Victoria*, 95, 147–168.

—— and Wasson, R. J. 1984 'Glacial age environments of arid Australia'. In J. C. Vögel (ed.), *Late Cainozoic Palaeoclimates of the Southern Hemisphere*, 183–208, A. A. Balkema, Rotterdam.

Breed, C. S. and Breed, W. J. 1979 'Dunes and other windforms of central Australia (and a comparison with linear dunes on the Moenkopi Plateau, Arizona)'. In F. El-Baz and D. M. Warner (eds), *Apollo-Soyuz Test Project vol. 2: Earth Observations and Photography*, National Technical Information Service, Washington DC.

——, Fryberger, S. G., Andrews, S., McCauley, C., Lennartz, F., Geber, D. and Horstman, K. 1979 'Regional studies of sand seas using LANDSAT (ERTS) imagery'. In E. D. McKee (ed.), *A Study of Global Sand Seas*, 305–398, United States Geological Survey, Professional Paper 1052.

—— and Grow, T. 1979 'Morphology and distribution of dunes in sand seas observed by remote sensing'. In E. D. McKee (ed.), *A Study of Global Sand Seas*, 253–304, United States Geological Survey, Professional Paper 1052.

——, McCauley, J. F. and Davis, P. A. 1987 'Sand sheets of the eastern Sahara and ripple blankets on Mars'. In L. E. Frostick and I. Reid (eds), *Desert Sediments: Ancient and Modern*, 337–359, Blackwell Scientific Publications, Oxford.

Brookfield, M. 1970 'Dune trends and wind regime in Central Australia'. *Zeitschrift für Geomorphologie*, Supplement, 10, 121–158.

Brookfield, M. E. 1977 'The origin of bounding surfaces in ancient aeolian sandstones'. *Sedimentology*, 24, 303–332.

REFERENCES

Brown, R. A. 1980 'Longitudinal instabilities and secondary flows in the planetary boundary layer: a review'. *Reviews of Geophysics and Space Physics*, 18, 683–697.

Burkinshaw, J. R., Illenberger, W. K. and Rust, I. C. 1993 'Wind speed profiles over a reversing transverse dune'. In K. Pye (ed.), *The Dynamics and Environmental Context of Aeolian Sedimentary Systems*, 25–36, Geological Society Special Publication 72, London.

Cailleux, A. 1952 'L'indice de emousée des grain de sable et grès'. *Revue de Géomorphologie Dynamique*, 2, 78–87.

Capot-Rey, R. 1947 'Dry and humid morphology in the western erg'. *Geographical Review*, 35, 391–407.

—— 1970 'Remarques sur les ergs du Sahara'. *Annales de Géographie*, 79, 2–19.

—— and Gremion, M. 1964 'Remarques sur quelques sables Sahariens'. *Travaux de l'Institut de Recherches Sahariennes*, 23, 153–163.

Chawla, S., Dhir, R. P. and Singhvi, A. K. 1991 'Thermoluminescence chronology of sand profiles in the Thar Desert and their implications'. *Quaternary Science Reviews*, 11, 25–32.

Chen, X. Y., Prescott, J. R. and Hutton, J. T. 1990 'Thermoluminescence dating on gypseous dunes of Lake Amadeus, central Australia'. *Australian Journal of Earth Sciences*, 37, 93–101.

Chepil, W. S. 1945a 'Dynamics of wind erosion: I. Nature of movement of soil by wind'. *Soil Science*, 60, 305–320.

—— 1945b 'Dynamics of wind erosion: II. Initiation of soil movement'. *Soil Science*, 60, 397–411.

—— and Woodruff, N. P. 1963 'The physics of wind erosion and its control'. *Advances in Agronomy*, 15, 211–302.

Chorley, R. J. and Kennedy, B. A. 1971 *Physical Geography: a systems approach*, Prentice-Hall, London, 370 pp.

Clarke, M. I., Lancaster, N. and Wintle, A. G. in press 'Infra-red stimulated luminescence dating of sands from the Cronese Basins, Mojave Desert'. *Geomorphology*.

Clemmensen, L. B. 1987 'Complex star dunes and associated aeolian bedforms, Hopeman Sandstone (Permo-Triassic), Moray Firth Basin, Scotland'. In L. E. Frostick and I. Reid (eds), *Desert Sediments: Ancient and Modern*, 213–231, Blackwell, Oxford.

—— 1989 'Preservation of interdraa and plinth deposits by the lateral migration of large linear draas (Lower Permian Yellow Sands, northeast England)'. *Sedimentary Geology*, 65, 139–151.

Clos-Arceduc, A. 1966 'Le rôle déterminant des ondes adriennes stationnaires dans la structure des ergs sahariens et les formes d'érosion avoisinantes'. *Comptes Rendus de l'Académie des Sciences, Paris*, 262D, 2673–2676.

—— 1967 'La direction des dunes et ses rapports avec celle du vent'. *Comptes Rendus de l'Académie des Sciences, Paris*, 264D, 1393–1396.

—— 1971 'Typologie des dunes vives'. *Travaux de la Institut de Géographie de Reims*, 6, 63–72.

Colletta, B. and Ortlieb, L. 1984 'Deformations of the middle and late Pleistocene deltaic deposits at the mouth of the Rio Colorado, northwestern Gulf of California'. In V. e.a. Malpica-Cruz (ed.), *Neotectonics and sea level variations in the Gulf of California area: a symposium*, 31–53, Universidad Nacional Autónoma de México, Instituto de Geologia, Mexico City.

Cooke, R. U. and Warren, A. 1973 *Geomorphology in Deserts*, Batsford, London, 374 pp.

Cooper, W. S. 1958 'Coastal sand dunes of Oregon and Washington'. *Geological*

REFERENCES

Society of America Memoir, 72, 167 pp.

Corbett, I. 1993 'The modern and ancient pattern of sandflow through the southern Namib deflation basin'. *International Association of Sedimentologists Special Publication,* 16, 45–60.

Cornish, V. 1897 'On the formation of sand-dunes'. *Geographical Journal,* 9, 278–302.

—— 1914 *Waves of Sand and Snow,* Fisher Unwin, London, 383 pp.

Crocker, R. L. 1946 'The soil and vegetation of the Simpson Desert and its borders'. *Transactions of the Royal Society of South Australia,* 70, 235–258.

Deacon, J. and Lancaster, N. 1988 *Late Quaternary Environments of Southern Africa,* Oxford University Press, Oxford, 220 pp.

Dubief, J. 1952 'Le vent et la déplacement du sable du Sahara'. *Travaux de la Institut de Recherches Sahariennes,* 8, 123–164.

Dyer, K. 1986 *Coastal and Estuarine Sediment Dynamics,* John Wiley, Chichester.

Edgett, K. S. and Lancaster, N. 1993 'Volcaniclastic aeolian dunes'. *Journal of Arid Environments,* 25, 271–297.

Edwards, S. R. 1993 'Luminescence dating of sand from the Kelso Dunes, California'. In K. Pye (ed.), *Dynamics and Environmental Context of Aeolian Sedimentary Systems,* 59–68, Geological Society of London, Special Publication 72, London.

El Baz, F. 1978 'The meaning of desert color in earth orbital photographs'. *Photogrammetric Engineering and Remote Sensing,* 44, 71–75.

Ellwood, J. M., Evans, P. D. and Wilson, I. G. 1975 'Small scale aeolian bedforms'. *Journal of Sedimentary Petrology,* 45, 554–561.

Embabi, N. S. 1982 'Barchans of the Kharga Depression'. In F. El Baz and T. A. Maxwell (eds), *Desert Landforms of Egypt: A Basis for Comparison with Mars,* 141–156, NASA, Washington DC.

Endrody-Younga, S. 1982 'Dispersion and translocation of dune specialist tenebrionids in the Namib area'. *Cimbebasia (A),* 5, 257–271.

Engel, P. 1981 'Length of flow separation over dunes'. *Proceedings of the American Society of Civil Engineers,* 107, 1133–1143.

Evans, J. R. 1961 'Falling and climbing sand dunes in the Cronese ("Cat") Mountain area, San Bernadino County, California'. *Journal of Geology,* 70, 107–113.

Fabre, J. and Petit-Maire, N. 1988 'Holocene climatic evolution at 22–23° N from two palaeolakes in the Taoudeni area (northern Mali)'. *Palaeogeography, Palaeoclimatology, Palaeoecology,* 65, 133–148.

Finkel, H. J. 1959 'The barchans of Southern Peru'. *Journal of Geology,* 67, 614–647.

Flint, R. F. and Bond, G. 1968 'Pleistocene sand ridges and pans in Western Rhodesia'. *Geological Society of America Bulletin,* 79, 299–314.

Folk, R. 1970 'Longitudinal dunes of the northwestern edge of the Simpson Desert, Northern Territory, Australia, 1. Geomorphology and grain size relationships'. *Sedimentology,* 16, 5–54.

—— 1971 'Genesis of longitudinal and oghurd dunes elucidated by rolling upon grease'. *Geological Society of America Bulletin,* 82, 3461–3468.

—— 1976a 'Reddening of desert sands: Simpson Desert, N.T., Australia.'. *Journal of Sedimentary Petrology,* 46, 604–615.

—— 1976b 'Rollers and ripples in sand, streams and sky: rhythmic alteration of transverse and longitudinal vortices in three orders'. *Sedimentology,* 23, 649–669.

—— 1978 'Angularity and silica coatings of Simpson Desert sand grains, Northern Territory, Australia'. *Journal of Sedimentary Petrology,* 48, 611–624.

—— and Ward, W. C. 1957 'Brazos River bar: a study in the significance of grain size parameters'. *Journal of Sedimentary Petrology,* 27, 3–26.

Forman, S. L. and Maat, P. 1990 'Stratigraphic evidence for late Quaternary dune activity near Hudson on the piedmont of northern Colorado'. *Geology*, 18, 745–748.

Frank, A. in press 'Model of airflow patterns on the lee side of eolian dunes'. *Geomorphology*.

—— and Kocurek, G. 1994a 'Effects of atmospheric conditions on wind profiles and eolian sand transport with an example from White Sands National Monument'. *Earth Surface Processes and Landforms*, 19, 735–74.

—— and Kocurek, G. in press 'Airflow up sand dunes: limitations of current understanding'. *Geomorphology*.

Fredsoe, J. 1982 'Shape and dimension of stationary dunes in rivers'. *Journal of Hydraulics of the American Society of Civil Engineers*, 108 (HY8), 932–947.

Friedman, G. M. 1962 'On sorting, sorting coefficients, and the log normality of the grain-size distribution of sandstones'. *Journal of Geology*, 70, 737–753.

Fryberger, S. G. 1979 'Dune forms and wind regimes'. In E. D. McKee (ed.), *A Study of Global Sand Seas*, 137–140, United States Geological Survey, Professional Paper 1052.

—— 1980 'Dune forms and wind regime, Mauritania, West Africa: implications for past climate'. *Palaeocology of Africa*, 12, 79–96.

—— 1991 'Unusual sedimentary structures in the Oregon coastal dunes'. *Journal of Arid Environments*, 21, 131–150.

—— and Ahlbrandt, T. S. 1979 'Mechanisms for the formation of aeolian sand seas'. *Zeitschrift für Geomorphologie*, 23, 440–460.

——, Ahlbrandt, T. and Andrews, S. 1979 'Origin, sedimentary features, and significance of low-angle eolian "sand sheet" deposits, Great Sand Dunes National Monument and vicinity, Colorado'. *Journal of Sedimentary Petrology*, 49, 733–746.

——, Al-Sari, A. M. and Clisham, T. J. 1983 'Eolian dune, interdune, sand sheet and siliciclastic sabkha sediments of an offshore prograding sand sea, Dhahran area, Saudi Arabia'. *American Association of Petroleum Geologists Bulletin*, 67, 280–312.

——, Al-Sari, A. M., Clisham, T. J., Rizoi, S. A. R. and Al-Hinai, K. G. 1984 'Wind sedimentation in the Jafarah sand sea, Saudi Arabia'. *Sedimentology*, 31, 413–431.

—— and Goudie, A. S. 1981 'Arid Geomorphology'. *Progress in Physical Geography*, 5, 420–428.

——, Hesp, P. and Hastings, K. 1992 'Aeolian granule ripple deposits, Namibia'. *Sedimentology*, 39, 319–331.

—— and Schenk, C. J. 1988 'Pin stripe lamination: a distinctive feature of modern and ancient eolian sediments'. *Sedimentary Geology*, 55, Special Issue: Eolian Sediments, ed. Hesp, P. and Fryberger, S. G., 1–15.

Gardner, G. J., Mortlock, A. J., Price, D. M., Readhead, M. L. and Wasson, R. J. 1987 'Thermoluminescence and radiocarbon dating of Australian desert dunes'. *Australian Journal of Earth Science*, 34, 343–357.

Gardner, R. 1988 'Aeolianites and marine deposits of the Wahiba Sands: character and palaeoenvironments'. In R. W. Dutton (ed.), *Scientific Results of the Royal Geographical Society's Oman Wahiba Sands Project 1985–1987*, 75–94, *Journal of Oman Studies*, Special Report 3, Muscat, Oman.

—— and Pye, K. 1981 'Nature, origin and paleoenvironmental significance of red coastal and desert dune sands'. *Progress in Physical Geography*, 5, 514–534.

Gautier, E. F. 1935 *The Great Sahara*, Columbia University Press, New York, 264 pp.

Gaylord, D. R. 1990 'Holocene paleoclimatic fluctuations revealed from dune and interdune strata in Wyoming'. *Journal of Arid Environments*, 18, 123–138.

REFERENCES

Gerety, K. M. 1985 'Problems with determination of u* from wind-velocity profiles measured in experiments with saltation'. In O. E. Barndorff-Nielsen, J. T. Møller, K. R. Rasmussen and B. B. Willetts (eds), *Proceedings of International Workshop on the Physics of Blown Sand*, 271–300, University of Aarhus, Aarhus.

—— and Slingerland, R. 1983 'Nature of the saltating population in wind tunnel experiments with heterogeneous size-density sands'. In M. E. Brookfield and T. S. Ahlbrandt (eds), *Eolian Sediments and Processes*, 115–131, Elsevier, Amsterdam.

Gillette, D. A., Adams, J., Muhs, D. and Kihl, R. 1982 'Threshold friction velocities and rupture moduli for crusted desert soils for the input of soil particles in the air'. *Journal of Geophysical Research*, 87, 9003–9015.

—— and Stockton, P. H. 1989 'The effect of nonerodible particles on the wind erosion of erodible surfaces'. *Journal of Geophysical Research*, 94, 12885–12893.

Glennie, K. W. 1970 *Desert Sedimentary Environments*, Elsevier, Amsterdam, 222.

Goudie, A. 1970 'Notes on some major dune types in Southern Africa'. *The South African Geographical Journal*, 52, 93–101.

—— and Thomas, D. S. G. 1985 'Pans in southern Africa with particular reference to South Africa and Zimbabwe'. *Zeitschrift für Geomorphologie*, 29, 1–19.

——, Warren, A., Jones, D. K. C. and Cooke, R. U. 1987 'The character and possible origins of the aeolian sediments of the Wahiba Sands'. *Geographical Journal*, 153, 231–256.

—— and Watson, A. 1981 'The shape of desert sand dune grains'. *Journal of Arid Environments*, 4, 185–190.

Greeley, R., Arvidson, R. E., Plaut, J. J., Saunders, R. S., Schubert, G., Sofan, E. R., Thouvenot, E. J. P., Wall, S. D. and Weitz, C. M. 1992 'Aeolian features on Venus: preliminary Magellan results'. *Journal of Geophysical Research*, 97, 13319–13345.

——, Blumberg, D. G. and Williams, S. H. in press 'Field measurements of the flux and speed of windblown sand'. *Sedimentology*.

——, Gaddis, L., Lancaster, N., Dobrovolskis, A., Iversen, J. D., Rasmussen, K. R., Saunders, R. S., Van Zyl, J., Wall, S., Zebker, H. and White, B. R. 1991 'Assessment of aerodynamic roughness via airborne radar observations'. *Acta Mechanica*, Supplement 2, 77–88.

—— and Iversen, J. D. 1985 *Wind as a Geological Process*, Cambridge University Press, Cambridge, 333 pp.

—— and Iversen, J. D. 1987 'Measurements of wind friction speeds over lava surfaces and assessment of sediment transport'. *Geophysical Research Letters*, 14, 925–928.

——, Iversen, J. D., Pollack, J. B., Udovich, N. and White, B. R. 1974 'Wind tunnel studies of martian aeolian processes'. *Proceedings of the Royal Society of London*, 341, 331–360.

——, Lancaster, N., Lee, S. and Thomas, P. 1992 'Martian aeolian processes, sediments and features'. In H. Kieffer, B. M. Jakosky, C. W. Snyder and M. S. Matthews (eds), *Mars*, 730–767, University of Arizona Press, Tucson.

——, White, B., Leach, R., Iversen, J. and Pollack, J. 1976 'Mars: wind friction speeds for particle movement'. *Geophysical Research Letters*, 3, 417.

Grove, A. T. 1969 'Landforms and climatic change in the Kalahari and Ngamiland'. *Geographical Journal*, 135, 190–212.

—— and Warren, A. 1968 'Quaternary landforms and climate on the south side of the Sahara'. *Geographical Journal*, 134, 189–208.

Gunatilaka, A. and Mwango, S. B. 1989 'Flow separation and the internal structure of shadow dunes'. *Sedimentary Geology*, 61, 125–134.

Hack, J. T. 1941 'Dunes of the Western Navajo County'. *Geographical Review*, 31, 240–263.

REFERENCES

Haff, P. K. and Presti, D. E. 1984 *Barchan Dunes of the Salton Sea Region, California*. California Institute of Technology, Department of Physics, Brown Bag Preprint Series, BB-16.

Hallet, B. 1990 'Spatial self-organization in geomorphology: from periodic bedforms and patterned ground to scale-invariant topography'. *Earth Science Reviews*, 29, 57–76.

Hanna, S. R. 1969 'The formation of longitudinal sand dunes by large helical eddies in the atmosphere'. *Journal of Applied Meteorology*, 8, 874–883.

Hardisty, J. and Whitehouse, R. J. S. 1988a 'Evidence for a new sand transport process from experiments on Saharan dunes'. *Nature*, 332, 532–534.

—— and Whitehouse, R. J. S. 1988b 'Effect of bedslope on desert sand transport'. *Nature*, 334, 302.

Hartmann, D. and Christiansen, C. 1988 'Settling velocity distributions and sorting processes on a longitudinal sand dune'. *Earth Surface Processes and Landforms*, 13, 649–656.

Hastenrath, S. L. 1967 'The barchans of the Arequipa region, Southern Peru'. *Zeitschrift für Geomorphologie*, 11, 300–311.

—— 1987 'The barchan dunes of Southern Peru revisited'. *Zeitschrift für Geomorphologie*, 31, 167–178.

Havholm, K. G. and Kocurek, G. 1988 'A preliminary study of the dynamics of a modern draa, Algodones, southeastern California, USA'. *Sedimentology*, 35, 649–669.

Haynes, C. V., Jr 1982 'Great Sand Sea and Selima Sand Sheet, Eastern Sahara: geochronology of desertification'. *Science*, 217, 629–633.

—— 1989 'Bagnold's barchan: a 57-yr record of dune movement in the eastern Sahara and implications for dune origin and palaeoclimate since Neolithic times'. *Quaternary Research*, 32, 153–167.

—— and Mead, A. R. 1987 'Radiocarbon dating and paleoclimatic significance of subfossil Limicolaria in northwestern Sudan'. *Quaternary Research*, 28, 86–99.

Hedin, S. 1905 'The Central Asian Deserts, sand dunes and sands'. In *Scientific Results of a Journey in Central Asia 1899–1902*, vol. 2, 379–718, Swedish Army Lithographic Institute, Stockholm.

Heine, K. 1982 'The main stages of the late Quaternary evolution of the Kalahari region, southern Africa'. *Palaeoecology of Africa*, 15, 53–76.

Hesp, P. A. 1981 'The formation of shadow dunes'. *Journal of Sedimentary Petrology*, 51, 101–112.

Holliday, V. T. 1989 'Middle Holocene drought on the southern High Plains'. *Quaternary Research*, 31, 74–82.

Holm, D. A. 1960 'Desert geomorphology in the Arabian Peninsula'. *Science*, 123, 1369–1379.

Hotta, S., Kubota, S., Katori, S. and Horikawa, K. 1984 'Sand transport by wind on a wet sand surface'. *Coastal Engineering–1984*, 1265–1281.

Howard, A. D. 1977 'Effect of slope on the threshold of motion and its application to orientation of wind ripples'. *Geological Society of America Bulletin*, 88, 853–856.

—— 1985 'Interaction of sand transport with topography and local winds in the northern Peruvian coastal desert'. In O. E. Barndorff-Nielsen, J. T. Møller, K. R. Rasmussen and B. B. Willetts (eds), *Proceedings of International Workshop on the Physics of Blown Sand*, Aarhus, University of Aarhus, 511–544.

——, Morton, J. B., Gad-el-Hak, M. and Pierce, D. B. 1978 'Sand transport model of barchan dune equilibrium'. *Sedimentology*, 25, 307–338.

Hoyt, J. 1965 'Air and sand movements to the lee of dunes'. *Sedimentology*, 7, 137–143.

REFERENCES

Hsu, S. 1971 'Wind stress criteria in eolian sand transport'. *Journal of Geophysical Research*, 76, 8684–8686.

Hummel, G. and Kocurek, G. 1984 'Interdune areas of the back-island dune field, North Padre Island, Texas'. *Sedimentary Geology*, 39, 1–26.

Hunt, J. C. R., Leibovich, S. and Richards, K. J. 1988 'Turbulent shear flows over low hills'. *Quarterly Journal of the Royal Meteorological Society*, 114, 1435–1470.

—— and Simpson, J. E. 1982 'Atmospheric boundary layers over non-homogeneous terrain'. In E. Plate (ed.), *Engineering Meteorology*, 269–318, Elsevier, Amsterdam.

Hunter, R. E. 1977 'Basic types of stratification in small eolian dunes'. *Sedimentology*, 24, 361–388.

—— 1985 'A kinematic model for the structure of lee-side deposits'. *Sedimentology*, 32, 409–422.

——, Richmond, B. M. and Alpha, T. R. 1983 'Storm-controlled oblique dunes of the Oregon Coast'. *Geological Society of America Bulletin*, 94, 1450–1465.

Huthnance, J. M. 1982 'On one mechanism forming linear sand banks'. *Estuarine, Coastal, and Shelf Science*, 14, 79–99.

Hyde, R. and Wasson, R. J. 1983 'Radiative and meteorological control on the movement of sand at Lake Mungo, N.S.W., Australia'. In M. E. Brookfield and T. S. Ahlbrandt (eds), *Eolian Sediments and Processes*, 311–323, Elsevier, Amsterdam.

Inman, D. L., Ewing, G. C. and Corliss, J. B. 1966 'Coastal sand dunes of Guerrero Negro, Baja California, Mexico'. *Geological Society of America Bulletin*, 77, 787–802.

Iversen, J. D., Wen-Ping Wang, Rasmussen, K. R., Mikkelsen, H., Hasiuk, J. F. and Leach, R. N. 1990 'The effect of a roughness element on local saltation transport'. *Journal of Wind Engineering and Industrial Aerodynamics*, 36, 845–854.

—— and White, B. R. 1982 'Saltation threshold on Earth, Mars and Venus'. *Sedimentology*, 29, 111–119.

Jackson, P. S. and Hunt, J. C. R. 1975 'Turbulent wind flow over a low hill'. *Quarterly Journal of the Royal Meteorological Society*, 101, 929–955.

Jackson II, R. G. 1975 'Hierarchcial attributes and a unifying model of bed forms composed of cohesionless material and produced by shearing flow'. *Geological Society of America Bulletin*, 86, 1523–1533.

Jäkel, D. 1980 'Die bildung von barchanen in Faya-Largeau/Rep. du Tchad'. *Zeitschrift für Geomorphologie*, 24, 141–159.

Jennings, J. N. 1968 'A revised map of the desert dunes of Australia'. *Australian Geographer*, 10, 408–409.

Johnson, R. B. 1967 'The Great Sand Dunes of southern Colorado'. *United States Geological Survey Professional Paper*, 575-C, 177–183.

Kaiser, E. 1926 *Die Diamantenwüste Südwestafrikas*. Dietrich Reimer, Berlin.

Kar, A. 1990 'The megabarchanoids of the Thar: their environment, morphology and relationship with longitudinal dunes'. *Geographical Journal*, 156, 51–61.

Kawamura, R. 1951 'Study of sand movement by wind'. *Institute of Science and Technology, Tokyo, Report*, 5, 3–4 Tokyo, Japan, 95–112.

Kelley, R. D. 1984 'Horizontal roll and boundary layer interrelationships observed over Lake Michigan'. *Journal of Atmospheric Science*, 41, 1816–1826.

Kennedy, J. F. 1969 'The formation of sediment ripples, dunes and antidunes'. *Annual Reviews of Fluid Mechanics*, 1, 147–169.

Khalaf, F. I., Al-Ajmi, D., Saleh, N. and Al-Hashash, M. 1993 'Aeolian processes and sand encroachment problems in Kuwait'. *Geomorphology*, 6, 111–134.

King, D. 1960 'The sand ridge deserts of South Australia and related aeolian

REFERENCES

landforms of Quaternary arid cycles'. *Transactions of the Royal Society of South Australia*, 79, 93–103.

Kocurek, G. 1981 'Significance of interdune deposits and bounding surfaces in eolian dune sands'. *Sedimentology*, 28, 753–780.

—— 1986 'Origins of low-angle stratification in aeolian deposits'. In W. G. Nickling (ed.), *Aeolian Geomorphology*, 177–193, Allen and Unwin, Boston.

—— 1988 'First order and super bounding surfaces in eolian sequences – Bounding surfaces revisited'. *Sedimentary Geology*, 56, 193–206.

—— and Dott, R. H. J. 1981 'Distinctions and uses of stratification types in the interpreation of eolian sand'. *Journal of Sedimentary Petrology*, 51, 579–595.

—— and Havholm, K. G. 1994 'Eolian sequence stratigraphy – a conceptual framework'. In P. Weimer and H. Posamentier (eds), *Siliciclastic Sequence Stratigraphy*, 393–409, American Association of Petroleum Geologists, Tulsa, Oklahoma.

——, Havholm, K. G., Deynoux, M. and Blakey, R. C. 1991 'Amalgamated accumulations resulting from climatic and eustatic changes, Akchar Erg, Mauritania'. *Sedimentology*, 38, 751–772.

—— and Nielson, J. 1986 'Conditions favourable for the formation of warm-climate aeolian sand sheets'. *Sedimentology*, 33, 795–816.

——, Townsley, M., Yeh, E., Havholm, K. and Sweet, M. L. 1992 'Dune and dunefield development on Padre Island, Texas, with implications for interdune deposition and water-table-controlled accumulation'. *Journal of Sedimentary Petrology*, 62, 622–635.

Krinsley, D. H. and Trusty, P. 1985 'Environmental interpretation of quartz grain surface textures'. In G. G. Zuffa (ed.), *Provenance of Arenites*, 213–229, Reidel, Dordrecht.

Lai, R. J. and Wu, J. 1978 *Wind Erosion and Deposition along a Coastal Sand Dune*. University of Delaware Sea Grant Program, Report DEL-SG-10-78.

Lancaster, N. 1978 'Composition and formation of southern Kalahari pan margin dunes'. *Zeitschrift für Geomorphologie*, 22, 148–169.

—— 1980 'The formation of seif dunes from barchans – supporting evidence for Bagnold's hypothesis from the Namib Desert'. *Zeitschrift für Geomorphologie*, 24, 160–167.

—— 1981a 'Palaeoenvironmental implications of fixed dune systems in southern Africa'. *Palaeogeography, Palaeoclimatology, Palaeoecology*, 33, 327–346.

—— 1981b 'Grain size characteristics of Namib Desert linear dunes'. *Sedimentology*, 28, 115–122.

—— 1982a 'Dunes on the Skeleton Coast, SWA/Namibia: geomorphology and grain size relationships'. *Earth Surface Processes and Landforms*, 7, 575–587.

—— 1982b 'Linear dunes'. *Progress in Physical Geography*, 6, 476–504.

—— 1983a 'Controls of dune morphology in the Namib sand sea'. In T. S. Ahlbrandt and M. E. Brookfield (eds), *Eolian Sediments and Processes*, 261–289, Elsevier, Amsterdam.

—— 1983b 'Linear dunes of the Namib sand sea'. *Zeitschrift für Geomorphologie*, Supplementband 45, 27–49.

—— 1984 'Aridity in southern Africa: age, origins and expression in landforms and sediments'. In J. C. Vogel (ed.), *Late Cenozoic Palaeoclimates of the Southern Hemisphere*, 433–444, A. A. Balkema, Rotterdam.

—— 1985a 'Winds and sand movements in the Namib sand sea'. *Earth Surface Processes and Landforms*, 10, 607–619.

—— 1985b 'Variations in wind velocity and sand transport rates on the windward flanks of desert sand dunes'. *Sedimentology*, 32, 581–593.

—— 1986 'Grain size characteristics of linear dunes in the southwestern Kalahari'. *Journal of Sedimentary Petrology*, 56, 395–400.

—— 1987 'Variations in wind velocity and sand transport rates on the windward flanks of desert sand dunes. Reply to comments by A. Watson'. *Sedimentology*, 34, 511–520.

—— 1988a 'Controls of eolian dune size and spacing'. *Geology*, 16, 972–975.

—— 1988b 'Development of linear dunes in the southwestern Kalahari, southern Africa'. *Journal of Arid Environments*, 14, 233–244.

—— 1988c 'The development of large aeolian bedforms'. *Sedimentary Geology*, 55, 69–89.

—— 1989a 'The dynamics of star dunes: an example from the Gran Desierto, Mexico'. *Sedimentology*, 36, 273–289.

—— 1989b *The Namib Sand Sea: Dune Forms, Processes, and Sediments*, A. A. Balkema, Rotterdam, 200 pp.

—— 1989c 'Late Quaternary Palaeoenvironments in the southwestern Kalahari'. *Palaeogeography, Palaeoclimatology, Palaeoecology*, 70, 367–376.

—— 1989d 'Star dunes', *Progress in Physical Geography*, 13, 67–92.

—— 1990 'Palaeoclimatic evidence from sand seas'. *Palaeogeography, Palaeoclimatology, Palaeoecology*, 76, 279–290.

—— 1991 'The orientation of dunes with respect to sand-transporting winds: a test of Rubin and Hunter's gross bedform-normal rule'. *Acta Mechanica*, Supplement 2, 89–102.

—— 1992 'Relations between dune generations in the Gran Desierto, Mexico'. *Sedimentology*, 39, 631–644.

—— 1993a 'Origins and sedimentary features of supersurfaces in the northwestern Gran Desierto Sand Sea'. *Aeolian Sediments: Ancient and Modern*, IAS Special Publication, 16, 71–86.

—— 1993b 'Development of Kelso Dunes, Mojave Desert, California'. *National Geographic Research and Exploration*, 9, 444–459.

—— 1994a 'Dune morphology and dynamics'. In A. D. Abrahams and A. J. Parsons (eds), *Geomorphology of Desert Environments*, 474–505, Chapman and Hall, London.

—— 1994b 'Controls on aeolian activity: new perspectives from the Kelso Dunes, Mojave Desert, California'. *Journal of Arid Environments*, 27, 113–124.

——, Greeley, R. and Christensen, P. R. 1987 'Dunes of the Gran Desierto Sand Sea, Sonora, Mexico'. *Earth Surface Processes and Landforms*, 12, 277–288.

—— and Nickling, W. G. 1994 'Aeolian sediment transport'. In A. D. Abrahams and A. J. Parsons (eds), *Geomorphology of Desert Environments*, 447–473, Chapman and Hall, London.

——, Nickling, W. G., McKenna Neuman, C. K. and Wyatt, V. E. 1994 'Sediment flux and airflow on the stoss slope of a barchan dune'. In N. Lancaster (ed.), *Response of Eolian Processes to Global Change: Abstracts of a workshop*, Desert Research Institute, Reno.

—— and Ollier, C. D. 1983 'Sources of sand for the Namib Sand Sea'. *Zeitschrift für Geomorphologie*, Supplementband 45, 71–83.

——, Rasmussen, K. R. and Greeley, R. 1991 'Interactions between unvegetated desert surfaces and the atmospheric boundary layer: a preliminary assessment'. *Acta Mechanica*, Supplement 2, 89–102.

—— and Teller, J. T. 1988 'Interdune deposits of the Namib Sand Sea'. *Sedimentary Geology* 55, 91–107.

Le Mone, M. A. 1973 'The structure and dynamics of horizontal roll vortices in the planetary boundary layer'. *Journal of Atmospheric Science*, 30, 1077–1091.

REFERENCES

Lee, J. A. 1991 'Near-surface wind flow around desert shrubs'. *Physical Geography*, 12, 140–146.

Lees, B. G., Yanchou, L. and Head, J. 1990 'Reconnaissance thermoluminescence dating of Northern Australian coastal dune systems'. *Quaternary Research*, 34, 169–185.

Lettau, K. and Lettau, H. H. 1978 'Experimental and micro-meteorological field studies on dune migration'. In K. Lettau and H. H. Lettau (eds), *Exploring the World's Driest Climate*, 110–147, University of Wisconsin, Madison, Institute for Environmental Studies.

Lewis, A. D. 1936 'Sand dunes of the Kalahari within the borders of the Union'. *South African Geographical Journal*, 14, 22–32.

Livingstone, I. 1986 'Geomorphological significance of wind flow patterns over a Namib linear dune'. In W. G. Nickling (ed.), *Aeolian Geomorphology*, 97–112, Boston, Allen and Unwin.

—— 1987a 'Photographic evidence of seasonal change in a secondary form on a "complex" linear dune'. *Madoqua*, 15, 237–241.

—— 1987b 'Grain-size variation on a "complex" linear dune in the Namib Desert'. In L. Frostick and I. Reid (eds), *Desert Sediments: Ancient and modern*, 281–291, Geological Society Special Publication No. 35.

—— 1988 'New models for the formation of linear sand dunes'. *Geography*, 73, 105–115.

—— 1989 'Monitoring surface change on a Namib linear dune'. *Earth Surface Processes and Landforms*, 14, 317–332.

—— 1993 'A decade of surface change on a Namib linear dune'. *Earth Surface Processes and Landforms*, 18, 661–664.

—— and Thomas, D. S. G. 1993 'Modes of linear dune activity and their significance: an evaluation with reference to southern African examples'. In K. Pye (ed.), *Dynamics and Environmental Context of Aeolian Sedimentary Systems*, 91–102, Geological Society of London, Special Publication 72, London.

Logan, R. F. 1960 *The Central Namib Desert, South West Africa*, National Academy of Sciences, Washington DC, 162 pp.

Logie, M. 1982 'Influence of roughness elements and soil moisture of sand to wind erosion'. *Catena Supplement*, 1, 161–173.

Long, J. T. and Sharp, R. P. 1964 'Barchan dune movement in Imperial Valley, California'. *Geological Society of America Bulletin*, 75, 149–156.

Loope, D. B. 1984 'Origin of extensive bedding planes in eolian sandstones: a defense of Stokes' hypothesis – discussion'. *Sedimentology*, 31, 123–125.

Lyles, L. and Krauss, R. K. 1971 'Threshold velocities and initial particle motion as influenced by air turbulence'. *Transactions of American Society of Agricultural Engineers*, 14, 563–566.

——, Schrandt, R. L. and Schneidler, N. F. 1974 'How aerodynamic roughness elements control sand movement'. *Transactions of American Society of Agricultural Engineers*, 17, 134–139.

Mabbutt, J. A. and Wooding, R. A. 1983 'Analysis of longitudinal dune patters in the northwestern Simpson Desert, central Australia'. *Zeitschrift für Geomorphologie*, Supplement 45, 51–69.

McClure, H. A. 1978 'Ar Rub' al Khali'. In S. S. Al-Sayari and J. G. Zötl (eds), *Quaternary Period in Saudi Arabia. 1: Sedimentological, Hydrochemical, Geomorphological Investigations in Central and Eastern Saudi Arabia*, 252–263, Springer-Verlag, Vienna, New York.

McCoy, F. W., Noeleberg, W. J. and Norris, R. M. 1967 'Speculations on the origin of the Algodones Dunes, California'. *Geological Society of America Bulletin*, 78, 1039–1044.

REFERENCES

McEwan, I. K. 1993 'Bagnold's knik: a physical feature of a wind velocity profile modified by blown sand?' *Earth Surface Processes and Landforms*, 18, 145–156.

—— and Willetts, B. B. 1991 'Numerical model of the saltation cloud'. *Acta Mechanica*, Supplement 1, 53–66.

—— and Willetts, B. B. 1993 'Sand transport by wind: a review of the current conceptual model'. In K. Pye (ed.), *Dynamics and Environmental Context of Aeolian Sedimentary Systems*, 7–16, Geological Society of London, Special Publication 72, London.

——, Willetts, B. B. and Rice, M. A. 1992 'The grain/bed collision in sand transport by wind'. *Sedimentology*, 39, 971–983.

McKee, E. D. 1966 'Structures of dunes at White Sands National Monument, New Mexico (and a comparison with structures of dunes from other selected areas)'. *Sedimentology*, 7, 1–69.

—— 1979a 'Introduction to a study of global sand seas'. In E. D. McKee (ed.), *A Study of Global Sand Seas*, 3–19, United States Geological Survey, Professional Paper 1052.

—— 1979b 'Sedimentary structures in dunes'. In E. D. McKee (ed.), *A Study of Global Sand Seas*, 83–136, United States Geological Survey, Professional Paper 1052.

—— 1982 *Sedimentary Structures in Dunes of the Namib Desert, South West Africa*. Geological Society of America Special Paper, 188, 60 pp.

—— and Douglass, J. R. 1971 *Growth and Movement of Dunes at White Sands National Monument, New Mexico*. United States Geological Survey.

——, Douglass, J. R. and Rittenhouse, S. 1971 'Deformation of lee-side laminae in eolian dunes'. *Geological Society of America Bulletin*, 82, 359–378.

—— and Moiola, R. J. 1975 'Geometry and growth of the White Sands Dune Field, New Mexico'. *United States Geological Survey Journal of Research*, 8, 59–66.

—— and Tibbitts, G. C., Jr 1964 'Primary structures of a seif dune and associated deposits in Libya'. *Journal of Sedimentary Petrology*, 34, 5–17.

McKenna-Neuman, C. and Nickling, W. G. 1989 'A theoretical and wind tunnel investigation of the effect of capillary water on the entrainment of sediment by wind'. *Canadian Journal of Soil Science*, 69, 79–96.

McLean, S. R. and Smith, J. D. 1986 'A model for flow over two-dimensional bedforms'. *Journal of Hydraulic Engineering*, 112, 300–317.

Madigan, C. T. 1946 'The Simpson Desert expedition, 1939, Scientific Reports: No. 6, Geology – the sand formations'. *Transactions of the Royal Society of South Australia*, 70, 45–151.

Madole, R. F. 1994 'Stratigraphic evidence of desertification in the west-central Great Plains within the past 1000 yr'. *Geology*, 22, 483–486.

Mainguet, M. 1972 'Etude d'un erg (Fachi-Bilma) son alimentation sableuse et sa insertion dans le paysage d'après les photographies prises par satellites'. *Comptes-Rendus de l'Académie des Sciences, Paris*, 274, 1633–1636.

—— 1983 'Dunes vives, dunes fixées, dunes vêtues: une classification selon le bilan d'alimentation, le régime éolien et la dynamique des édifices sableux'. *Zeitschrift für Geomorphologie*, Supplement Bd. 45, 265–285.

—— 1984a 'A classification of dunes based on aeolian dynamics and the sand budget'. In F. El Baz (ed.), *Deserts and Arid Lands*, 31–58, Martinus Nijhoff, The Hague.

—— 1984b 'Space observations of Saharan aeolian dynamics'. In F. El Baz (ed.), *Deserts and Arid Lands*, 59–77, Nyjhoff, The Hague.

—— 1985 'Le Sahel, un laboratoire naturel pour l'étude du vent, mécanisme principal de la désertification', *Proceedings of International Workshop on the*

REFERENCES

Physics of Blown Sand, Aarhus, University of Aarhus, Denmark, 545–563.

────── and Callot, Y. 1978 'L'erg de Fachi-Bilma (Tchad-Niger)'. *Mémoires et Documents CNRS*, 18, 178.

────── and Canon, L. 1976 'Vents et paléovents du Sahara: tentative d'approache paléoclimatique'. *Revue de Géographie Physique et de Géologie Dynamique*, 18, 241–250.

────── and Chemin, M.-C. 1983 'Sand seas of the Sahara and Sahel: an explanation of their thickness and sand dune type by the sand budget principle'. In M. E. Brookfield and T. S. Ahlbrandt (ed.), *Eolian Sediments and Processes*, 353–364, Elsevier, Amsterdam.

────── and Chemin, M.-C. 1984 'Les dunes pyramidales du Grand Erg Oriental'. *Travaux de l'Institut de Géographie de Reims*, 59–60, 49–60.

────── and Jacqueminet, C. 1984 'Le Grand Erg Occidental et le Grand Erg Oriental'. *Travaux de l'Institut de Géographie de Reims*, 59–60, 29–48.

Marrs, R. W. and Kolm, K. E. 1982 'Interpretation of eolian processes and windflow patterns from eolian landforms: An introduction'. In R. W. Marrs and K. E. Kolm (eds), *Interpretation of Windflow Characteristics from Eolian Landforms*, Geological Society of America Special Paper 192, 1–3, Geological Society of America, Boulder, Colorado.

Marshall, J. K. 1971 'Drag measurements in roughness arrays of varying densities and distribution'. *Agricultural Meteorology*, 8, 269–292.

Mason, C. C. and Folk, R. L. 1958 'Differentiation of beach, dune, and aeolian flat environments by size analysis, Mustang Island, Texas'. *Journal of Sedimentary Petrology*, 28, 211–226.

Matschinski, M. 1952 'Sur les formations sableuses des environs de Beni-Abbes'. *Compte Rendu de la Societé géologique de France*, 9–10, 171–174.

Maxwell, T. A. and Haynes, C. V., Jr 1989 'Large-scale, low-amplitude bedforms (chevrons) in the Selima sand sheet, Egypt'. *Science*, 243, 1179–1182.

Merriam, R. 1969 'Source of sand dunes of southeastern California and northwestern Sonora, Mexico'. *Geological Society of America Bulletin*, 80, 531–534.

Middleton, N. J., Goudie, A. S. and Wells, G. L. 1986 'The frequency and source areas of dust storms'. In W. G. Nickling (ed.), *Aeolian Geomorphology*, 237–259, Allen and Unwin, London.

Middleton, G. V. and Southard, J. B. 1984 *Mechanics of Sediment Movement*, Society of Economic Palaeontologists and Mineralogists, Tulsa, Oklahoma, 401.

Monod, T. 1958 *Majabat al-Koubra*, Mémoires de L'Institut Français d'Afrique Noire, 52, 395 pp.

Muhs, D. R. 1985 'Age and palaeoclimatic significance of Holocene san dunes in northeastern Colorado'. *Association of American Geographers Annals*, 75, 566–582.

────── in press 'Source of sand for the Algodones Dunes'. In V. P. Tchakerian (ed.), *Aeolian Processes in Deserts*, Chapman and Hall, New York.

────── and Maat, P. B. 1993 'The potential response of eolian sands to Greenhouse Warming and precipitation reduction on the Great Plains of the United States'. *Journal of Arid Environments*, 25, 351–361.

Mulligan, K. R. 1988 'Velocity Profiles measured on the windward slope of a transverse dune'. *Earth Surface Processes and Landforms*, 13, 573–582.

Musick, H. B. and Gillette, D. A. 1990 'Field evaluation of relationships between a vegetation structural parameter and sheltering against wind erosion'. *Land Degradation and Rehabilitation*, 2, 87–94.

Nagtegaal, P. J. C. 1971 'Adhesion-ripple and barchan-dune sands of the recent Namib (SW Africa) and Permian Rotliegend (NW Europe) Deserts'. *Madoqua*, Ser. II, Vol. 2, 5–19.

REFERENCES

Nanson, G. C., Chen, X. Y. and Price, D. M. 1992 'Lateral migration, thermoluminescence chronology, and colour variation of longitudinal dunes near Birdsville in the Simpson Desert, Australia'. *Earth Surface Processes and Landforms*, 17, 807–820.

———, Price, D. M. and Short, S. A. 1992 'Wetting and drying of Australia over the past 300 ka'. *Geology*, 20, 791–794.

Nelson, J. M. and Smith, J. D. 1989 'Mechanics of flow over ripples and dunes'. *Journal of Geophysical Research*, 75 (C6), 8146–8162.

Newbold, D. 1924 'A desert odyssey of a thousand miles'. *Sudan Notes and Records*, 7, 43–92, 104–107.

Nickling, W. G. 1983 'Grain-size characteristics of sediment transported during dust storms'. *Journal of Sedimentary Petrology*, 53, 1011–1024.

——— 1984 'The stabilizing role of bonding agents on the entrainment of sediment by wind'. *Sedimentology*, 31, 111–117.

——— 1988 'Initiation of particle movement by wind'. *Sedimentology*, 35, 499–512.

——— and Ecclestone, M. 1981 'The effects of soluble salts on the threshold shear velocity of fine sand'. *Sedimentology*, 28, 505–510.

——— and Gillies, J. A. 1993 'Dust emission and transport in Mali, West Africa'. *Sedimentology*, 40, 859–868.

——— and McKenna-Neuman, C. K. 1995 'Development of deflation lag surfaces'. *Sedimentology*, 42.

——— and Wolfe, S. A. 1994 'The morphology and origin of nabkhas, region of Mopti, Mali, West Africa'. *Journal of Arid Environments*, 28, 13–30.

Nielson, J. and Kocurek, G. 1986 'Climbing zibars of the Algodones'. *Sedimentary Geology*, 48, 1–15.

——— and Kocurek, G. 1987 'Surface processes, deposits, and development of star dunes: Dumont dune field, California'. *Geological Society of America Bulletin*, 99, 177–186.

Norris, R. M. and Norris, K. S. 1961 'Algodones dunes of southeastern California'. *Geological Society of America Bulletin*, 72, 605–620.

Norstrud, H. 1982 'Wind flow over low arbitrary hills'. *Boundary-Layer Meteorology*, 23, 115–124.

O'Brien, M. P. and Rindlaub, B. D. 1936 'The transportation of sand by wind'. *Civil Engineering*, 6, 325–327.

Oke, T. R. 1978 *Boundary Layer Climates*, Methuen, New York, 372 pp.

Owen, P. R. 1964 'Saltation of uniform grains in air'. *Journal of Fluid Mechanics*, 20, 225–242.

Paisley, E. C. I., Lancaster, N., Gaddis, L. and Greeley, R. 1991 'Discrimination of active and inactive sands by remote sensing: Kelso Dunes, Mojave Desert, California'. *Remote Sensing of Environment*, 37, 153–166.

Partridge, T. C. 1993 'The evidence for Cainozoic aridification in southern Africa'. *Quaternary International*, 17, 105–110.

Passarge, S. 1904 *Die Kalahari*, Reimer, Berlin, 822 pp.

Pearse, J. R., Lindley, D. and Stevenson, D. C. 1981 'Wind flow over ridges in simulated atmospheric boundary layers'. *Boundary-Layer Meteorology*, 21, 77–92.

Petit-Maire, N., Casta, L., Delibrias, G., Gaven, C. and Testud, A.-M. 1980 'Preliminary data on Quaternary palaeolacustrine deposits in the Wadi ash Shati area, Libya'. In M. Salem and M. Busrewil (eds), *The Geology of Libya*, 797–807, Academic Press, London.

———, Fabre, J., Carbonel, P., Schulz, E. and Aucour, A. M. 1987 'La dépression de Taoudenni (Sahara malien) à l'Holocène'. *Géodynamique 2*, 2, 127–160.

Porter, M. L. 1986 'Sedimentary record of erg migration'. *Geology*, 14, 497–500.

REFERENCES

Prandtl, L. 1935 'The mechanics of viscous fluids'. In F. Durand (ed.), *Aerodynamic Theory Vol. III*, 57–10, Springer, Berlin.

Pye, K. 1987 *Aeolian Dust and Dust Deposits*, Academic Press, London, 334 pp.

—— and Lancaster, N. 1993 (eds) *Aeolian Sediments: Ancient and Modern*, IAS Special Publication 16, Blackwell, Oxford, 167.

—— and Tsoar, H. 1990 *Aeolian Sand and Sand Dunes*, Unwin Hyman, London, 396.

Rasmussen, K. R. and Mikkelsen, H. E. 1991 'Wind tunnel observations of aeolian transport rates'. *Acta Mechanica Supplementum*, 1, 135–144.

Rendell, H. in press 'Luminescence dating of sand ramps in the eastern Mojave Desert'. *Geomorphology*.

——, Lancaster, N. and Tchakerian, V. P. in press 'Luminescence dating of Late Pleistocene Aeolian Deposition in the Mojave Desert'. *Quaternary Geochronology*.

——, Yair, A. and Tsoar, M. 1993 'Thermoluminescence dating of periods of sand movement and linear dune formation in the northern Negev, Israel'. In K. Pye (ed.), *Dynamics and Environmental Context of Aeolian Sedimentary Systems*, 49–58, Geological Society of London, Special Publication 72, London.

Ritchie, J. C., Eyles, C. H. and Haynes, C. V. 1985 'Sediment and pollen evidence for an early to mid-Holocene humid period in the eastern Sahara'. *Nature*, 314, 352–354.

Rognon, P. 1982 'Pluvial and arid phases in the Sahara: the role of nonclimatic factors'. *Palaeoecology of Africa*, 12, 45–62.

—— 1987 'Late Quaternary climatic reconstruction for the Maghreb (North Africa)'. *Palaeogeography, Palaeoclimatology, Palaeoecology*, 58, 11–34.

—— and Williams, M. A. J. 1977 'Late Quaternary climatic changes in Australia and North Africa: a preliminary interpretation'. *Palaeogeography, Palaeoclimatology, Palaeoecology*, 21, 285–287.

Rubin, D. M. 1984 'Factors determining desert dune type (discussion)'. *Nature*, 309, 91–92.

—— and Hunter, R. E. 1982 'Bedform climbing in theory and nature'. *Sedimentology*, 29, 121–138.

—— and Hunter, R. E. 1987 'Bedform alignment in directionally varying flows'. *Science*, 237, 276–278.

—— and Ikeda, H. 1990 'Flume experiments on the alignment of transverse, oblique and longitudinal dunes in directionally varying flows'. *Sedimentology*, 37, 673–684.

—— and McCulloch, D. S. 1980 'Single and superimposed bedforms: a synthesis of San Francisco Bay and flume observations'. *Sedimentary Geology*, 26, 207–231.

Rumpel, D. A. 1985 'Successive aeolian saltation: studies of idealized collisions'. *Sedimentology*, 32, 267–280.

Sarnthein, M. 1978 'Sand deserts during glacial maximum and climatic optima'. *Nature*, 272, 43–46.

Sarre, R. D. 1989 'Aeolian sand transport'. *Progress in Physical Geography*, 11, 157–182.

Schenk, C. J., Gautier, D. L., Olhoeft, G. R. and Lucius, J. E. 1993 'Internal structure of an aeolian dune using ground-penetrating radar'. IAS Special Publication, 61–70, Blackwell, Oxford.

Schlicting, H. 1936 'Experimentelle untersuchungen zum rauhigkeitsproblem'. *Ingenieur-Archiv*, 7, 1–34.

Schumm, S. A. and Lichty, R. W. 1965 'Time, space and causality in geomorphology'. *American Journal of Science*, 263, 110–119.

REFERENCES

Seely, M. K. and Sandelowsky, B. H. 1974 'Dating the regression of a river's end point'. *South African Archaeological Bulletin*, Goodwin Series, 2, 61–64.

Seppälä, M. and Linde, K. 1978 'Wind tunnel studies of ripple formation'. *Geografiska Annaler*, 60, 29–42.

Sharp, R. P. 1963 'Wind Ripples'. *Journal of Geology*, 71, 617–636.

—— 1964 'Wind-driven sand in Coachella Valley, California'. *Geological Society of America Bulletin*, 75, 785–804.

—— 1966 'Kelso Dunes, Mohave Desert, California'. *Geological Society of America Bulletin*, 77, 1045–1074.

—— 1979 'Intradune flats of the Algodones chain, Imperial Valley, California'. *Geological Society of America Bulletin*, 90, 908–916.

Sherman, D. J. 1990 'Evaluation of aeolian sand transport equations using intertidal-zone measurements, Saunton Sands, England. Discussion'. *Sedimentology*, 37, 385–388.

Shinn, E. A. 1973 'Sedimentary accumulation along the leeward southeast coast of Qatar Peninsula, Persian Gulf'. In B. H. Purser (ed.), *The Persian Gulf*, 199–209, Springer-Verlag, New York.

Shotton, F. W. 1937 'The Lower Bunter sandstones of north Worcestershire and east Shropshire (England)'. *Geological Magazine*, 74, 534–553.

Simons, F. S. 1956 'A note on Pur-Pur Dune, Viru Valley, Peru'. *Journal of Geology*, 64, 517–521.

Simpson, E. and Loope, D. 1985 'Amalgamated interdune deposits, White Sands, New Mexico'. *Journal of Sedimentary Petrology*, 55, 0361–0365.

Slattery, M. C. 1990 'Barchan migration on the Kuiseb River Delta, Namibia'. *South African Geographical Journal*, 72, 5–10.

Smith, H. T. U. 1954 'Eolian sand on desert mountains'. *Geological Society of America, Annual Meeting Abstracts*, 102–103.

—— 1967 *Past versus Present Wind Action in the Mojave Desert Region, California*. US Army Cambridge Research Laboratory, Report AFCRL-67-0683.

Smith, J. D. and McLean, S. R. 1977 'Spatially averaged flow over a wavy surface'. *Journal of Geophysical Research*, 82, 1735–1746.

Smith, R. S. U. 1978 'Field trip to dunes at Superstition Mountain'. In R. Greeley, R. P. Papson and P. D. Spudis (eds), *Aeolian Features of Southern California: a comparative planetary geology guidebook*, 65–72, NASA.

—— 1980 'Maintenance of barchan size in the southern Algodones Dune Chain, Imperial County, California'. *Reports of the Planetary Geology Program*, NASA Technical Memorandum, 81776, 253–254.

—— 1984 'Eolian geomorphology of the Devils Playground, Kelso Dunes and Silurian Valley, California'. In J. C. Dohrenwend (ed.), *Surficial Geology of the eastern Mojave Desert, California*, 162–173, Geological Society of America, Boulder, Colorado.

Sorensen, M. 1985 'Estimation of some aeolian saltation transport parameters from transport rate profiles'. In O. E. Barndorff-Nielsen, J. T. Møller, K. R. Rasmussen and B. B. Willetts (eds), *Proceedings of the International Workshop on the Physics of Blown Sand*, University of Aarhus, Aarhus, Denmark, 141–190.

Spaulding, W. G. 1991 'A Middle Holocene vegetation record from the Mojave Desert of North America and its palaeoclimatic significance'. *Quaternary Research*, 35, 427–437.

Sprigg, R. C. 1979 'Stranded and submerged sea-beach systems of southeast south Australia and the aeolian desert cycle'. *Sedimentary Geology*, 22, 53–96.

—— 1982 'Alternating wind cycles of the Quaternary era and their influence in and around the dune deserts of southeastern Australia'. In R. J. Wasson (ed.),

REFERENCES

Quaternary Dust Mantles of China, New Zealand, and Australia, 211–240, ANU Press, Canberra.

Stockton, P. H. and Gillette, D. A. 1990 'Field measurements of the sheltering effect of vegetation on erodible land surfaces'. *Land Degradation and Rehabilitation*, 2, 77–86.

Stokes, S. 1991 'Luminescence dating of some archaeologically significant dune and fluvial deposits from the Llano Estacado'. *Society for American Archaeology*, Abstracts, 56th Annual Meeting, 159.

—— and Breed, C. S. 1993 'A chronostratigraphic re-evaluation of the Tusayan Dunes, Moenkopi Plateau and Ward Terrace, Northeastern Arizona'. In K. Pye (ed.), *The Dynamics and Environmental Context of Aeolian Sedimentary Systems*, 75–90, Geological Society Special Paper 72, London.

Stokes, W. L. 1968 'Multiple parallel-truncation bedding planes – a feature of wind-deposited sandstone'. *Journal of Sedimentary Petrology*, 55, 361–365.

Street, F. A. and Grove, A. T. 1979 'Global maps of lake level fluctuations since 30,000 yr B.P.'. *Quaternary Research*, 12, 83–118.

Sweet, M. L. 1989 'Eolian dune airflow dynamics: implications for dune migration, deposits, and spacing', unpublished Ph.D., University of Texas at Austin.

—— and Kocurek, G. 1990 'An empirical model of aeolian dune lee-face airflow'. *Sedimentology*, 37, 1023–1038.

——, Nielson, J., Havholm, K. and Farralley, J. 1988 'Algodones dune field of southeastern California: case history of a migrating modern dune field'. *Sedimentology*, 35, 939–952.

Talbot, M. R. 1980 'Environmental responses to climatic change in the West African Sahel over the past 20,000 years'. In M. A. J. Williams and H. Faure (eds), *The Sahara and the Nile*, 37–62, A. A. Balkema, Rotterdam.

—— 1984 'Late Pleistocene dune building and rainfall in the Sahel'. *Palaeoecology of Africa*, 16, 203–214.

—— 1985 'Major bounding surfaces in aeolian sandstones: a climatic model'. *Sedimentology*, 32, 257–266.

—— and Williams, M. A. J. 1978 'Erosion of fixed dunes in the Sahel, central Niger'. *Earth Surface Processes*, 3, 107–113.

Taylor, P. A., Mason, P. J. and Bradley, E. F. 1987 'Boundary-layer flow over low hills'. *Boundary-Layer Meteorology*, 39, 107–132.

Tchakerian, V. P. 1991 'Late Quaternary aeolian geomorphology of the Dale Lake sand sheet, southern Mojave Desert, California'. *Physical Geography*, 12, 347–369.

Teller, J. T. and Lancaster, N. 1986 'Lacustrine sediments at Narabeb in the central Namib Desert, Namibia'. *Palaeogeography, Palaeoclimatology, Palaeoecology*, 56, 177–195.

——, Rutter, N. W. and Lancaster, N. 1990 'Sedimentology and Palaeohydrology of Late Quaternary Lake Deposits in the Northern Namib Sand Sea, Namibia'. *Quaternary Science Reviews*, 9, 343–364.

Thesiger, W. 1949 'A further journey across the Empty Quarter'. *Geographical Journal*, 113, 21–46.

Thomas, D. S. G. 1984 'Ancient ergs of the former arid zones of Zimbabwe, Zambia, and Angola'. *Transactions of the Institute of British Geographers*, n.s. 9, 75–88.

—— 1986 'Dune pattern statistics applied to the Kalahari Dune Desert, Southern Africa'. *Zeitschrift für Geomorphologie*, 30, 231–242.

—— 1988 'Analysis of linear dune sediment–form relationships in the Kalahari dune desert'. *Earth Surface Processes and Landforms*, 13, 545–554.

—— 1989 (ed.) *Arid Zone Geomorphology*, Bellhaven/Halsted Press, London, 372 pp.

REFERENCES

—— 1992 'Desert dune activity: concepts and significance'. *Journal of Arid Environments*, 22, 31–38.

—— and Shaw, P. A. 1991a *The Kalahari Environment*, Cambridge University Press, Cambridge, 284.

—— and Shaw, P. A. 1991b '"Relict" desert dune systems: interpretations and problems'. *Journal of Arid Environments*, 20, 1–14.

—— and Tsoar, H. 1990 'The geomorphological role of vegetation in desert dune systems'. In J. B. Thornes (ed.), *Vegetation and Erosion*, 471–489, John Wiley, Chichester.

Thompson, R. W. 1968 'Tidal flat sedimentation on the Colorado River Delta, northwestern Gulf of California'. *Geological Society of America Memoir*, 125 pp.

Tseo, G. 1990 'Reconnaissance of the dynamic characteristics of an active Strzelecki Desert longitudinal dune, southcentral Australia'. *Zeitschrift für Geomorphologie N.F.*, 34, 19–35.

—— 1993 'Two types of longitudinal dune fields and possible mechanisms for their development'. *Earth Surface Processes and Landforms*, 18, 627–644.

Tsoar, H. 1974 'Desert dunes morphology and dynamics, El Arish (northern Sinai)'. *Zeitschrift für Geomorphologie Supplementband*, 20, 41–61.

—— 1978 *The Dynamics of Longitudinal Dunes. Final Technical Report to the US Army European Research Office*, US Army European Research Office, London.

—— 1982 'Internal structure and surface geometry of longitudinal (seif) dunes'. *Journal of Sedimentary Petrology*, 52, 0823–0831.

—— 1983a 'Dynamic processes acting on a longitudinal (seif) dune'. *Sedimentology*, 30, 567–578.

—— 1983b 'Wind tunnel modeling of echo and climbing dunes'. In M. E. Brookfield and T. S. Ahlbrandt (eds), *Eolian Sediments and Processes*, 247–259, Elsevier, Amsterdam.

—— 1984 'The formation of seif dunes from barchans – a discussion'. *Zeitschrift für Geomorphologie*, 28, 99–103.

—— 1985 'Profile analysis of sand dunes and their steady state significance'. *Geografiska Annaler*, 67A, 47–59.

—— 1986 'Two-dimensional analysis of dune profile and the effect of grain size on sand dune morphology'. In F. El Baz and M. H. A. Hassan (eds), *Physics of Desertification*, 94–108, Martinus Nyjhoff, Dordrecht.

—— 1989 'Linear dunes – forms and formation'. *Progress in Physical Geography*, 13, 507–528.

—— 1990 'The ecological background, deterioration and reclamation of desert dune sand'. *Agriculture, Ecosystems and Environment*, 33, 147–190.

—— and Møller, J. T. 1986 'The role of vegetation in the formation of linear sand dunes'. In W. G. Nickling (ed.), *Aeolian Geomorphology*, 75–95, Allen and Unwin, London.

—— and Pye, K. 1987 'Dust transport and the question of desert loess formation'. *Sedimentology*, 34, 139–154.

Twidale, C. R. 1972 'Evolution of sand dunes in the Simpson Desert, central Australia'. *Transactions of the Institute of British Geographers*, 56, 77–110.

—— 1980 'The Simpson Desert, central Australia'. *South African Geograhical Journal*, 62, 3–17.

Ungar, J. E. and Haff, P. K. 1987 'Steady-state saltation in air'. *Sedimentology*, 34, 289–299.

van Andel, T. H. 1964 'Recent marine sediments of the Gulf of California'. In T. H. van Andel and G. C. Shor (eds), *Marine geology of the Gulf of California*, 216–310, American Association of Petroleum Geologists, Tulsa, Oklahoma.

REFERENCES

Verstappen, H. T. 1968 'On the origin of longitudinal (seif) dunes'. *Zeitschrift für Geomorphologie N.F.*, 12, 200–220.

Vincent, P. J. 1984 'Particle size variation over a transverse dune in the Nafud as Sirr, central Saudi Arabia'. *Journal of Arid Environments*, 7, 329–336.

—— 1986 'Differentiation of modern beach and coastal dune sands – a logistic regression approach using the parameters of the hyperbolic function'. *Sedimentary Geology*, 49, 167–176.

—— 1988 'The response diagram and sand mixtures'. *Zeitschrift für Geomorphologie*, 32, 221–226.

Visher, G. S. 1969 'Grain size distributions and depositional processes'. *Journal of Sedimentary Petrology*, 39, 1074–1106.

Vogel, J. C. 1989 'Evidence of past climatic change in the Namib Desert'. *Palaeogeography, Palaeoclimatology, Palaeoecology*, 70, 355–366.

Vökel, J. and Grunert, J. 1990 'To the problem of dune formation and dune weathering during the Late Pleistocene and Holocene in the southern Sahara and Sahel'. *Zeitschrift für Geomorphologie*, 34, 1–17.

Walker, A. S., Olsen, J. W. and Bagen 1987 'The Badain Jaran Desert: remote sensing investigations'. *Geographical Journal*, 153, 205–210.

Walker, D. J. 1981 'An experimental study of wind ripples', unpublished M.Sc., Massachusetts Institute of Technology.

Walker, T. R. 1979 'Red color in dune sands'. In E. D. McKee (ed.), *A Study of Global Sand Seas*, 61–82, United States Geological Survey, Professional Paper 1052.

Walmsley, J. L. and Howard, A. D. 1985 'Application of a boundary-layer model to flow over an eolian dune'. *Journal of Geophysical Research*, 90, 10, 631–10, 640.

Warren, A. 1969 'A bibliography of desert dunes and phenomena'. In W. S. McGinnies and B. J. Goldman (eds), *Arid Lands in Perspective*, 75–99, University of Arizona Press, Tucson.

—— 1970 'Dune trends and their implications in the central Sudan'. *Zeitschift für Geomorphology*, Supplement 10, 154–180.

—— 1972 'Observations on dunes and bimodal sands in the Tenere desert'. *Sedimentology*, 19, 37–44.

—— 1988 'The dunes of the Wahiba Sands'. In R. W. Dutton (ed.), *Scientific Results of the Royal Geographical Society's Oman Wahiba Sands Project 1985–1987*, 131–160, *Journal of Oman Studies*, Special Report 3, Muscat, Oman.

—— and Knott, P. 1983 'Desert dunes: a short review of needs in desert dune research and a recent study of micro-meteorological dune initiation mechanisms'. In M. E. Brookfield and T. S. Ahlbrandt (eds), *Eolian Sediments and Processes*, 343–352, Elsevier, Amsterdam.

Wasson, R. J. 1983 'Dune sediment types, sand colour, sediment provenance and hydrology in the Strzelecki-Simpson Dunefield, Australia'. In M. E. Brookfield and T. S. Ahlbrandt (eds), *Eolian Sediments and Processes*, 165–195, Elsevier, Amsterdam.

—— 1984 'Late Quaternary environments in the desert dunefields of Australia'. In J. C. Vögel (ed.), *Late Cainozoic Palaeoclimates of the Southern Hemisphere*, 419–432, A. A. Balkema, Rotterdam.

——, Fitchett, K., Mackey, B. and Hyde, R. 1988 'Large-scale patterns of dune type, spacing, and orientation in the Australian continental dunefield'. *Australian Geographer*, 19, 89–104.

—— and Hyde, R. 1983a 'A test of granulometric control of desert dune geometry'. *Earth Surface Processes and Landforms*, 8, 301–312.

—— and Hyde, R. 1983b 'Factors determining desert dune type'. *Nature*, 304, 337–339.

REFERENCES

———— and Nanninga, P. M. 1986 'Estimating wind transport of sand on vegetated surfaces'. *Earth Surface Processes and Landforms*, 11, 505–514.

————, Rajaguru, S. N., Misra, V. N., Agrawal, D. P., Dhir, R. P., Singhvi, A. K. and Kameswara Rao, K. 1983 'Geomorphology, late Quaternary stratigraphy and paleoclimatology of the Thar dunefield'. *Zeitschrift für Geomorphologie*, Supplement, 45, 117–151.

Waters, M. R. 1983 'Late Holocene lacustrine chronology and archaeology of ancient Lake Cahuilla, California'. *Quaternary Research*, 19, 373–387.

Watson, A. 1986 'Grain-size variations on a longitudinal dune and a barchan dune'. *Sedimentary Geology*, 46, 49–66.

———— 1987 'Variations in wind velocity and sand transport rates on the windward flanks of desert sand dunes: comment'. *Sedimentology*, 34, 511–516.

———— 1990 'The control of blowing sand and mobile desert dunes'. In A. S. Goudie (ed.), *Techniques for Desert Reclamation*, 35–85, John Wiley, Chichester.

Wells, G. L. 1983 'Late-Glacial circulation over Central North America revealed by aeolian features', in F. A. Street-Perrott, M. Bevan and R. Ratcliffe (eds) *Variations in the Global Water Budget*, 317–330, D. Reidel, Dordrecht.

Wells, S. G., McFadden, L. D. and Schultz, J. D. 1990 'Eolian landscape evolution and soil formation in the Chaco dune field, southern Colorado Plateau, New Mexico'. In P. L. K. Knuepfer and L. D. McFadden (eds), *Soils and Landscape Evolution. Proceedings of the 21st Binghamton Symposium in Geomorphology*, 517–546, Amsterdam, Elsevier.

Weng, W. S., Hunt, J. C. R., Carruthers, D. J., Warren, A., Wiggs, G. F. S., Livingstone, A. and Castro, I. 1991 'Air flow and sand transport over sand dunes'. *Acta Mechanica Supplement*, 2, 1–22.

Werner, B. T. 1988 'A steady-state model of wind-blown sand transport'. *Journal of Geology*, 98, 1–17.

White, B. R. 1979 'Soil transport by winds on Mars'. *Journal of Geophysical Reseach*, 84, 4643–5651.

———— 1982 'Two-phase measurements of saltating turbulent boundary layer flow'. *International Journal of Multiphase Flow*, 9, 459–473.

———— and Schultz, J. C. 1977 'Magnus effect in saltation'. *Journal of Fluid Mechanics*, 81, 497–512.

Whitney, J. W., Faulkender, D. J. and Rubin, M. 1983 *The Environmental History and Present Condition of the Northern Sand Seas of Saudi Arabia*. United States Geological Survey, Open File Report OF-03-95.

Wiegand, J. P. 1977 'Dune morphology and sedimentology at Great Sand Dunes National Monument', unpublished MS, Colorado State University.

Wiggs, G. F. S. 1993 'Desert dune dynamics and the evaluation of shear velocity: an integrated approach'. In K. Pye (ed.), *The Dynamics and Environmental Context of Aeolian Sedimentary Systems*, 37–48, Geological Society of London, Special Paper 72, London.

Willetts, B. B. 1983 'Transport by wind of granular materials of different grain shapes and densities'. *Sedimentology*, 30, 669–679.

————, McEwan, I. K. and Rice, M. A. 1991 'Initiation of motion of quartz sand grains'. *Acta Mechanica*, Supplement 1, 123–134.

———— and Rice, M. A. 1986 'Collision in aeolian transport: the saltation/creep link'. In W. G. Nickling (ed.), *Aeolian Geomorphology*, 1–18, Allen and Unwin, London.

———— and Rice, M. A. 1988 'Effect of bedslope on desert sand transport'. *Nature*, 334, 302.

————, Rice, M. A. and Swaine, S. E. 1982 'Shape effects in aeolian grain transport'.

Sedimentology, 29, 409–417.

Williams, G. 1964 'Some aspects of the eolian saltation load'. *Sedimentology,* 3, 257–287.

Williams, J. J., Butterfield, G. R. and Clark, D. G. 1990 'Aerodynamic entrainment thresholds on impervious and permeable beds'. *Earth Surface Processes and Landforms,* 15, 255–264.

Williams, J. J., Butterfield, G. R. and Clark, D. G. 1994 'Aerodynamic entrainment threshold: effects of boundary layer flow conditions'. *Sedimentology,* 41, 309–328.

Wilson, I. G. 1971 'Desert sandflow basins and a model for the development of ergs'. *Geographical Journal,* 137, 180–199.

——— 1972 'Aeolian bedforms – their development and origins'. *Sedimentology,* 19, 173–210.

——— 1973 'Ergs'. *Sedimentary Geology,* 10, 77–106.

Wintle, A. G. 1993 'Luminescence dating of aeolian sands – an overview'. In K. Pye (ed.), *Dynamics and Environmental Context of Aeolian Sedimentary Systems,* 49–58, Geological Society of London, Special Paper 72, London.

——— and Huntley, D. J. 1982 'Thermoluminescence dating of sediments'. *Quaternary Science Reviews,* 1, 31–53.

———, Lancaster, N. and Edwards, S. R. 1994 'Infrared stimulated luminescence (IRSL) dating of late-Holocene aeolian sands in the Mojave Desert, California, USA'. *The Holocene,* 4, 74–78.

Wolfe, S. A. 1993 'Sparse vegetation as a surface control on wind erosion', unpublished Ph.D., University of Guelph, Canada.

——— and Nickling, W. G. 1993 'The protective role of sparse vegetation in wind erosion'. *Progress in Physical Geography,* 17, 50–68.

Wopfner, A. and Twidale, G. R. 1967 'Geomorphological history of the Lake Eyre basin'. In J. N. Jennings and J. A. Mabbutt (eds), *Landform Studies from Australia and New Guinea,* Cambridge University Press, Cambridge.

Yair, A. 1990 'Runoff generation in a sandy area'. *Earth Surface Processes and Landforms,* 15, 597–610.

Yalin, M. S. 1972 *Mechanics of Sediment Transport,* Pergamon Press, Oxford.

Zimbelman, J. R. and Williams, S. H. 1990 'Interbasin transport of aeolian sand, Mojave Desert, California'. *EOS,* 71, 1245.

INDEX

77; regional–scale systems of linear dunes 6; relict dune systems 233; sand seas 1, 198, 200; dune types 213–15; southern 249; wind strength 240, 243
avalanche deposits 85, 87–8, 119
avalanche face 51
avalanching, on lee side 139
Awbari sand sea, Libya, complex crescentic dunes 59

Badain Jaran sand sea, China, star dunes 73
Bagnold, R. A. 4, 7, 23, 31, 52, 64, 85, 93, 106, 132, 169; dune initiation 125–6; on entrainment processes 17; focal point kink in wind profiles 24–6; on helical roll vortex models 186–7; saltation and wind ripple path lengths 39, 42; sand transport equations 22, 27–8, 30, 34, 43, 136; threshold equation 22
Bahia del Adair 251
barchanoid dunes 45
barchanoid ridges 50, 54, 56–7
barchans 6, 7, 50, 51, 52–3, 124; migration rate and height 148–9; modification into linear dunes 169–70; morphology of 7; wind regimes 52, 53, 168
basalt 88
Basin Province, USA 71, 206
beaches 88
Beadnell, H. J. L. 4
bedform climbing 202–3
bedform hierarchies 48–50, 83, 193–4
bedform nucleation 123, 127
bedform reconstitution time 194, 220
Bellair, P. 114
Besler, H. 111, 191
bioturbation 100
Blandford, W. T. 4
Blumberg, D. G. 125
bonding agents, and threshold velocity 22
Botswana 235
boundary layer characteristics, effects on sand transport 11–15, 187
bounding surfaces, model for the formation of 87, 88
'bouquets de silk' 64
Bowler, J. M. 243

Breed, C. S. 73, 93, 212, 234
Breed, W. J. 93
brink 51

[14]C dating 231
California, southern 240; see also Algodones Dunes; Coachella Valley; Kelso Dunes
Callot, Y. 64, 212
Capot-Rey, R. 4, 114
carbonate sand 88
Cat Dune, Mojave Desert, falling dune 82, 83
Chepil, W.S. 23
Chott Djerid basin 245
cinder cones 123
cinders 88
climatic changes 204, 225–6, 256, 257; effects on dune development 252–4; effects on sand sea development 245–52; responses of dunes to 228, 229–30; and sediment budgets 201
climbing dunes 48, 82, 83
climbing translatent strata 85, 86
Clos-Arceduc, A. 188
Coachella Valley 27, 240
Colorado, eastern, parabolic dunes 76; see also Great Sand Dunes
Colorado River 225, 251
complex dunes 48, 49; development of 190–3; size and spacing 194–5, 197
compound dunes 48, 49; development of 190–3; size and spacing 194–5, 197
computer simulations: of saltation 19; of sediment surfaces 40
Cooke, R. U. 186
Cooper, W. S. 127, 139
coppice dunes 77
cores 252
Cornish, V. 4, 188, 189
crescentic dunes 45, 46, 50–9; barchans 52–3; complex 50, 59; compound 50, 57–9; morphometry of 58; wind regimes in 59; erosion and deposition patterns on, 146–9; grain size and sorting characteristics 105, 106; relations between height and spacing 56; sedimentary structures 93, 94, 95; simple, morphometry of 56; spacing 189–90; and wind regimes 53, 166, 168

Lightning Source UK Ltd.
Milton Keynes UK
UKOW031214230613

212673UK00002B/13

9 780415 060943